RADIO FREQUENCY IDENTIFICATION FROM SYSTEM TO APPLICATIONS

RADIO FREQUENCY IDENTIFICATION FROM SYSTEM TO APPLICATIONS

Editor

Izaan Ahmed

scitus academics

Radio Frequency Identification from System to Applications
Edited by **Izaan Ahmed**

ISBN: 978-1-68117-602-4
Library of Congress Control Number: 2016946457

SCITUS Academics LLC,
616, Corporate Way, Suite 2, 4766,
Valley Cottage, NY 10989

www.scitusacademics.com

Notice

Printed in United States of America on Acid Free Paper

PREFACE

In many service industries, purchasing and distribution logistics, industry, manufacturing companies and material flow systems, automated identification procedures (Auto ID) have become extremely popular recently. There are automatic identification procedures for providing information on transit people, animals, products and goods. The ubiquitous barcode labels that caused a revolution in identification systems a long time ago are now found to be insufficient in a growing number of instances.

Presenting the latest developments and state-of-the-art results in the RFID field to address these challenges; the book offers a systematic and comprehensive description of the methodologies, architectures and technologies of different efficient, scalable, secure and reliable applications based on RFID and RFID. Identification of radio frequency (RFID) would be the use of radio waves to read and capture information stored on a tag placed on an object. A tag can be checked from a distance of up to several feet and must not be in the direct view of the viewer to be tracked. Technological developments in the field of RFID technology are taking place at such a rate that, although a book describes common scientific principles in this way, it is not sufficiently powerful to be able to examine the latest trends in the latest products on the market and the latest regulations and standards. It is also increasingly difficult not to lose track of applications with the extensive use of RFID technology. In ever shorter intervals, this book will provide information about new RFID applications. The book is intended for practitioners who want or may need to get as detailed an overview of the many technologies, the legal framework or even possible applications of RFID.

TABLE OF CONTENTS

7

RFID APPLICATIONS AND CHALLENGES

Ming-Shen Jian[1] and Jain-Shing Wu[2]

[1] Deptartment of Computer Science and Information Engineering, National Formosa University, Taiwan, R. O. C.

[2] Deptartment of Computer Science and Engineering, National Sun Yat-sen University, Taiwan, R. O. C.

1. INTRODUCTION

Radio Frequency identification (RFID) is the popular wireless induction system [1-7]. The same as the general bar code identification, each RFID tag in an RFID system is assumed that equips a unique ID (UID) itself. A standard RFID system is consisted of Tag, Reader, and Application. When an independent RFID tag approaches the RFID antenna, the induction between RFID tag and antenna happens. The RFID antenna reads the information and content recorded in the tag. Then the information is translated into the computational data by the RFID reader. Due to the portable RFID tag and untouched data transmission, many local or small area wireless applications for track and trace based on RFID systems were proposed.

RFID today is the popular wireless induction system [2-3,8-11]. Each RFID tag in RFID system is given a unique ID (UID) which records the on demand information. When an independent RFID tag approaches the RFID antenna, the induction between RFID tag and antenna happens. The information and content recorded in the tag is transmitted to the RFID antenna and translated into the computational data. Following up the data translation, the tag recognition can be completed and related applications are provided.

The RFID applications about agriculture that are now in widespread use such as the Animal Identification, the Product Record, and the Manufacturing process management. First, users identify products or materials via the tag and the reader of RFID system, and followed that recorded the data of products or materials on database in foregoing proposals. In a subsequent process, an especial function chose some suitable data from databases for analyzing, integration and description. This process helps users to understand the position and situation about products or materials. Therefore, by using the RFID, the main contributions in agriculture can be described as follows:

1) The data of manufacturing record was transformed from artificial process to electrical process.

2) Increased management (such as Inventory or Supply Chain) and production planning and scheduling efficiency than before.

3) Reduced the cost of operation and increased the economic effect.

4) Supply product safety information to customers to refer.

Many tracking applications based on ubiquitous computing and communication technologies have been presented in recent years such as RFID systems [4,5,7]. Therefore, RFID can be used to trace objects and asset worldwide. In addition, some warehouse systems or supply chain management systems can be combined with RFID to form goods tracking systems. The tracking systems help enterprises to manage their raw materials and products that reduce the cost of operation budget. However, more and more applications of RFID system that were introduced by people, and that the agriculture is the one of them.

Due to the popularity of RFID, many local or small area wireless applications were also proposed. The RFID tags were proposed to be used in hospital or health care [12-15]. Patients should always wear the RFID tag is designed for identification. The patient's current location and condition is monitored every time and everywhere within the hospital. It means that patients are under cared even an emergency state happens. Some entrance guard systems are also based on RFID system. The RFID ticket or RFID card [2, 3, 8, 10] is used to identify that a user is legal or not. According to the short-distance wireless signal, the RFID tag users can be monitored within the specific area. In other words, the RFID systems are generally used to be the hardware identification in many applications. In opposition to using the RFID system as the hardware identification, many software applications adopt software encryption as the identifications to protect the intellectual property of the applications or files. Considering the serious situations of pirate, intellectual property protection is important and becomes a famous issue.

Password protection is the popular encryption method to protect the applications. Each application or file of software is assigned an on demand given serial numbers or calculation function. People who use this application have to input the correct serial number then enable the application. Considering today's applications, personal multimedia services or software applications are popular. Customers use the personal multimedia devices such as MP3, PDA, iPod, Laptop, etc., to download the multimedia or application files from the server or website via Internet. In other words, many files or data are disseminated and exchanged via Internet. In addition, many hackers can crash the software encryption with fewer costs (Only program tools or applications needed). It makes that the piratical files are transmitted widely and the protection of intellectual property exists in name only.

For the purpose that the right of intellectual property and the right of the valid users are further protected and maintained, integration of the software and hardware encryption is needed. Since each RFID tag with a unique ID (UID) which records the on demand information can be used as the individual identification, the small and cheap RFID tag can be considered as the hardware/software encryption/decryption key corresponding to the files or applications. In the next section, we give some descriptions for related RFID application and system.

2. RELATED APPLICATION AND SYSTEM

Some researchers presented that the embedding RFID can be plugged into a small device such as handheld host [1]. The handheld device users can plug in the SD or CF interface of reader card. Hence, the users can scan and induct the RFID tag everywhere. In other words, to integrate the RFID system hardware into the mobile devices is practicable. Furthermore, the RFID system including RFID induction antenna, RFID parser and reader, RFID tag, etc., today is cheap. In addition, the RFID hardware including antenna and reader is not only cheap but also can be a PnP device. It means that the RFID hardware can be used as a normal user device such as the card-reader.

2.1. Rfid Encryption and Decryption For Intellectual Property Protection

2.1.1. Application

Since the RFID systems are popular and ripe for distinguishing treatment of individual target [16, 17], the unique characteristic or identification of RFID can be the solution of intellectual property protection. Many researches proposed the possible way to protect the intellectual property, products, or applications. In some applications [18], the RFID chips are embedded in the cap of bottle. The medicine can be differentiated between fake and true. In addition, the RFID chip can be placed in the CD or DVD disk. The CD-ROM can accesses and reads the information of the RFID for valid identification check. Only the CD or DVD with the authorized RFID can be played. Although the content is protected, the self-made content that burned in the CD-R/RW or DVD-R/RW may not provide the authorized RFID information. In other words, the private, non-business, or free digital content made by the individual may be limited and cannot be transmitted free. In addition, even the CD or DVD disks are protected, the digital content such as files or data still can be copied from the disk to other devices such as hard disc or MP3 player. Therefore, how to separate the right of the digital content for each user and how to protect the digital content from illegal use become the important issues.

Due to the demand of existed system integration, some applications related to *RFID Encryption and Decryption for Intellectual Property Protection* includes: PnP Middleware, RFID Hardware, End User RFID Device and End User RFID Tag, and Encryption/Decryption Procedure. The system framework is shown as follows.

For a normal user, there are two types of RFID devices for the encryption/decryption on RFID system (E/DonRFID system): End User RFID Device for digital content or multimedia information gaining, and End User RFID Tag for indentifying the legal user.

E/DonRFID not only provides the RFID based protection procedure but also includes the Encryption/Decryption method based on RFID character. The encryption and decryption can be implemented by hardware or software solution. The original digital data is encrypted by 1) hardware, 2) software, or 3) combination of hardware and software. Corresponding to the encryption method, suitable RFID tag of user for decrypting is needed.

Since three possible ways to protect the digital content are proposed, for the end users, there will be at least three possible states and method of *E/DonRFID Encryption/Decryption*, to gain the protected digital data, shown as follows:

1) Encryption and Decryption by Hardware and Software combination,

2) Encryption only by Hardware with Hardware and Software combination Decryption

3) Encryption only by Software with Hardware and Software combination Decryption

4) Encryption only by Hardware with Hardware Decryption

5) Encryption only by Software with Hardware Decryption

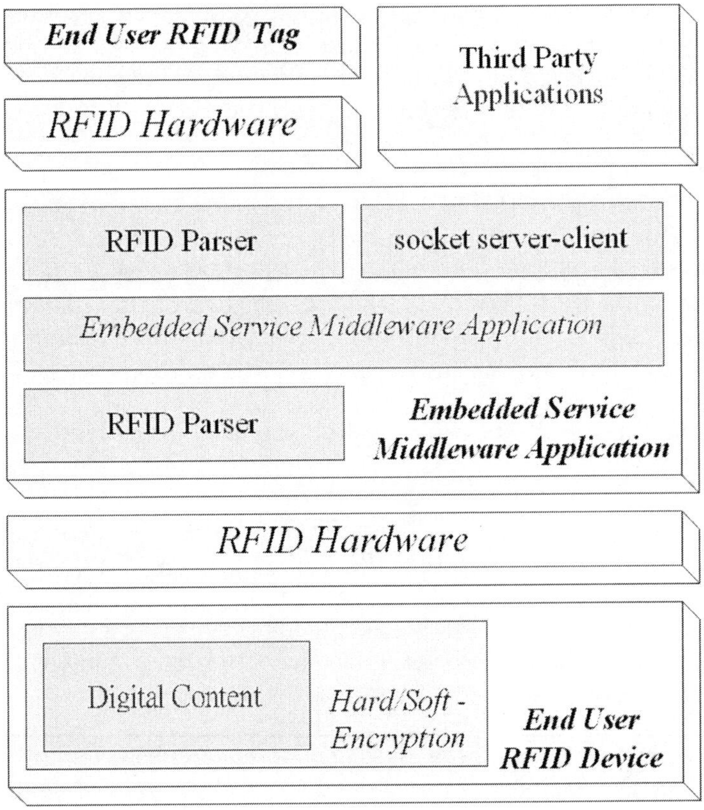

Figure 1. The framework of RFID Encryption and Decryption for Intellectual Property Protection

2.1.2. Method

First, depending on the digital content storage hardware such as CD-ROM disk, the commercial RFID tag can be embedded into the disk when the disk is made. According to the characteristic of RFID tag, each RFID tag can be set with different individualities. The different encryption code, unique ID, information of the digital content, or authentication serial number can be recorded in the RFID tag. In addition, the RFID tag embedded in the disk is not rewritable. Hence, different disks equip the different IDs of RFID tag. When the RFID reader inducts the tag, the information about this storage can be scanned and presented. In other words, only the digital storage with the valid RFID tag is legal and true.

Second, since the content or data are digital, these software, content or data, can be encrypted as the secret codes or cipher. The key for encryption and decryption can be recorded in the RFID tag. Without the specific key, these secret codes or ciphers cannot be recovered as the original data. In other words, the digital content that recorded in the storage device (such as CD-ROM disk) can be secured. The decryption key can be recorded in the RFID tag embedded in the storage or a palm RFID tag (such as a RFID toy).

For the end users, *End User RFID Device/Tag* is used. The storage, whether hardware (CD-ROM) which includes the encrypted digital content, or software (files or ciphers), is called *End User RFID Device.* If the *End User RFID Device* is hardware, the third party *RFID Hardware* can induct the RFID tag embedded in the hardware. After identifying the *End User RFID Device*, the application or user can execute and read the digital content if only *Hardware - Encryption/Decryption* is used.

According to three possible states, the end user must have the decryption key for executing the digital content. In this paper, the hardware (RFID tag) or software for the decryption key is called *End User RFID Tag*. After identifying the *End User RFID Device*, the end user has to provide the *End User RFID Tag* for the *Embedded Service Middleware Application*. Only the information or password of*End User RFID Tag* is correct and can be used to gain the secured decryption key which recorded in the *End User RFID Device*, the digital content recorded in the *End User RFID Device* can be presented.

Considering that the three possible states are based on the RFID induction, the *RFID Hardware* is divided into two types of equipments: for *End User RFID Device* and for *End User RFID Tag*.

According to the three possible ways to protect the digital content, when the protection is based on the combination of *Hard/Soft-Encryption/Decryption* and Only *Hardware-Encryption* with *Hard/Soft – Decryption*, the *RFID Hardware* for *End User RFID Device* is needed. Due to that the digital content is protected by the RFID tag embedded in the hardware, the information recorded in the tag has to be inducted before using. For example, if a tag is embedded in the CD-ROM disk, the user should have a CD-ROM driver with the *RFID Hardware* when reading the disk. In other words, if the protection is based on the hardware belongs to *End User RFID Device*, the corresponding reader with *RFID Hardware* is necessary. The *RFID Hardware* can be embedded in the CD-ROM driver, reader, or other multimedia devices.

In opposition to *End User RFID Device*, when the decryption is based on the *End User RFID Tag* key, end user has to own the valid RFID tag for decrypting the digital content. For example, the decryption code is recorded in the RFID tag of *End User RFID Device*. However, the decryption code is secured by the password which locks the data slot of RFID tag. Without the correct password, end user cannot gain the decryption code that secured in the RFID tag of *End User RFID Device*. To provide the password, the end users should have the *RFID Hardware* such as the USB-RFID reader, etc.

To manage the RFID information, *Embedded Service Middleware Application* is proposed to parse the information from the *RFID Hardware*. Due to that there are different RFID product, an RFID parser is needed for analyzing and parsing the information from *RFID Hardware*. After gaining the requirements or response, the *Embedded Service Middleware Application* searches the corresponding applications and passes the information. Figure 2 presents the framework of *Embedded Service Middleware Application*.

The *Embedded Service Middleware Application* implements the socket server-client structure for communication with other existed or third party applications. The information comes from the *End User RFID Device*, such as specific password-requirement, will be recorded in the database of middleware application. The requirement will be maintained based on the on demand limitation of the period of validity or when the *End User RFID Device* is removed. In addition, when an end user tries to gain the digital data from the *End User RFID Device*, the middleware application request the end user for the password. After receiving the password, the middle application transmits the password and tries to gain the decryption code. If the password is correct, the decryption code will be transmitted to the user application such as multimedia player, etc. Otherwise, the digital content cannot be decrypted and used. Therefore, only the two conditions: 1) the key information of *End User RFID Tag* matches the password requirement of *End User RFID Device*, and 2) the decryption code is correct in decrypting the digital content are satisfied, the user can gain the information from the *End User RFID Device*

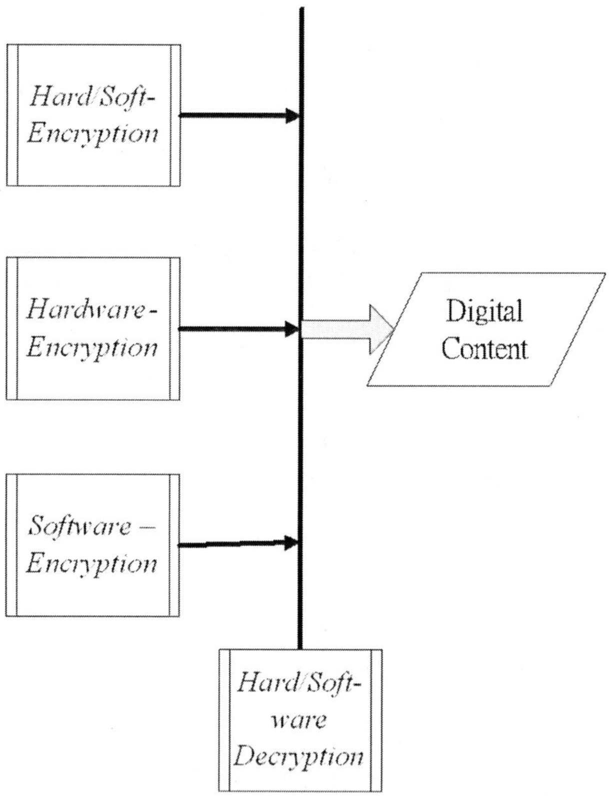

Figure 2. Possible ways to gain the protected digital data

2.2. Location Aware Public/Personal Information Services Based On Embedded Rfid Platform

2.2.1. Application

Many researches proposed before presented the importance of providing information and services related the user's location to each person. Some researches assume that there are GPS devices or module included in the users' mobile devices. Then, according to the information of GPS (GIS) [19-22], the location aware or related information or services are provided to the mobile user trough the wireless network. Although the GPS provides the accurate location of users, most users indeed needs the approximate local-area-aware information. The accuracy of location such as longitude and latitude is not the main issue. Furthermore, not everyone can equip the GPS.

Hence, in addition to GPS, according to the orientation made by the station of wireless cellular system [23], the related information according to the user's location can be given to the user via cellular system. Each cellular phone user can be served directly by the telecommunications companies. If a user is served by the specific wireless base station, the information related to the coverage area of this base station are given to the phone user.

Since the RFID system is popular and generally implemented, many researches [24-30] tried to integrate the RFID to and applied RFID technology to context-aware systems. However, in [31], what kind of the context, the corresponding context services, and the context-aware RFID system are important to be provided for user is still an issue of the existing system. In addition, to integrate the existed system such as information service and payment system become the important topic.

Not only supply the public services but also give the personal services, the context aware researches [32-34] were also proposed. Research in [35] was proposed that considering the user's related location. Hospital or health care RFID systems [12-14,34] for monitor the tag users were also proposed. A designed RFID tag is given to each user such as a patient. Each patient should always wear the RFID tag every time and everywhere. Hence, the patients' current information such as location and health conditions are monitored by the hospital. In addition, some entrance guard systems are also based on RFID system. The RFID ticket or RFID card [2, 3, 10, 36] is used to identify that a user is legal or not.

The services and information of user-location-related public places such as the museum [37] are provided. According to the requirement of users, different services are given through wireless network or cellular system to different users even they are in the same places.

Hence, a realistic application such as Location Aware Public/Personal Information Services based on RFID is proposed. By using the location-aware RFID application, the main contributions are :

1) Users of location-aware RFID application can communicate and gain the information corresponding to the users' location through the RFID tag. The handheld devices with RFID reader can also manually obtain the extra or required or local information and services.

2) The efficiency of system management and service utilization can be improved, information can be the digital multimedia and updated immediately,

3) The location-aware RFID application can be embedded in other similar service systems and hardware. The proposed service system can be included in the existed information center or server. The additional cost for integration can be reduced.

4) The service object and function can be various. For example, the location-aware RFID application provides not only the public or general information services to every system user, but also the deferential personal service to individual location-aware RFID application user.

5) The ticket and payment services can be integrated into location-aware RFID application service system. Users needn't to bring too many identification devices or cards. All method of payment and public or personal services are integrated into one RFID tag and location-aware RFID application service system.

2.2.2. Method

The system structure is shown as Figure 2. The *Embedded Service Middleware Platform* is the main system to manage the internal and external system connections. The RFID API and parser are included and provided to communicate with the third party RFID system. The *Embedded Service Middleware Platform* also makes the information connection to other business management system or database via software API. In addition, the related information to the RFID tag inducted is presented by user interface.

For the end users, *End User RFID Handheld Facilities* consists of two appliances: *end user RFID tag*and *end user device with RFID System.* A user can use a given readable and re-writable RFID tag or a handheld device such as PDA which equipped a RFID system to gain the required public/personal services. In Figure 2, the user handheld device also equips the RFID system, RFID API, and parser to scan and induct the commercial RFID tag. The communication and the data transmission between the handheld device and server can be established via 1) Internet, 2) server-client socket, 3) a user RFID tag, or 4) a readable and re-writable RFID tag. A user can view the information or obtain the services via user interface (UI) presented by server or the user handheld device.

The other business management systems in the framework can be the third party developments and independent of the whole system. When the user approaches the RFID system at the specified area, the induction and communication between end user RFID tag and antenna of *RFID System* is automatically established. A RFID reader will parse the signal into the digital and computing content. Then, the *RFID System* transmits the information obtained from the tag to the *Embedded Service Middleware Platform* via Internet. According to the RFID information, the *Embedded Service Middleware Platform* searches for and provides the specific personal service recorded in local area server according to the on demand conditions of the user. Moreover, the information or services can be updated or provided from the main database via Internet connection. Then, the user can obtain the public/personal information from the user interface.

In the location-aware RFID application service system, the RFID antennas and reader are deployed 1) at the specific area or location such as the entrance of the rapid transit system or the information service machine, or 2) within the handheld devices such as PDA or mobile phone. When a user is given a readable RFID tag, the related information or the user's on demand service conditions about the user is given by himself and on demand recorded in the database. When the user requires the local area public or personal services, the user should be at the tiny induction area such as a local area information center or a service station. Then, the RFID system placed in the specific area inducts the RFID tag and gain the information such as UID from the RFID tag. The reader of RFID system then sends the information to the local area server via Internet.

After receiving the information and parsing the message from RFID system, the content of RFID tag can be identified. If *end user RFID tag* is used, the *embedded server service middleware* can search and present the local information such as local area shopping information, traffic information, or the customization information, recorded in local database that match the on demand conditions of the RFID tag user. In other words, the RFID user can be directly served with sufficient local area related information. If other further information needed, the *embedded server service middleware* can send the user's request to the remote main server to obtain the requested service or to the other business applications via Internet for extra service obtaining. At last, the user can gain the location-aware information or services via user interface. In opposition to *end user RFID tag*, when a user of *end user device with RFID System* actively scans the RFID tag of the commercial advertisement, the handheld device can send the scanned RFID tag information via wireless network or cellular mobile system to the local area server with *embedded server service middleware* embedded. Then, as the procedure of*end user RFID tag*, the *embedded server service middleware* searches for the requested services and transmits these services to the user's handheld device by wireless network or cellular mobile system.

In addition, the users can use *handheld device middleware application* to select the tag content recorded in handheld device database if needed. Then, the RFID API controls the RFID system embedded in the handheld device to re-write the content (such as UID) of the tag of the handheld device. At last, the RFID content requirement from other business applications or systems can be provided through the RFID tag of the handheld device.

In addition, the database can record the history of the user's requirements. The statistic user requirements can be used to classify that what kind of the service the user requests most. Next time the system can provide the personal services according to the classified results. In other words, the users can be served with the services they most pay attention to.

The real test and verification is implemented as Figure 3. The implementation shows one shoot of the verification. When the user approaches the on demand placed RFID system, the *Embedded Service Middleware Platform* automatically presents the information corresponding the content or user's related information recorded in the RFID tag. For example, if a Taiwanese uses the RFID tag, the presentation of local area server will be based on traditional Chinese. But the local area server will provide English when a native English speaker user his own RFID tag respectively.

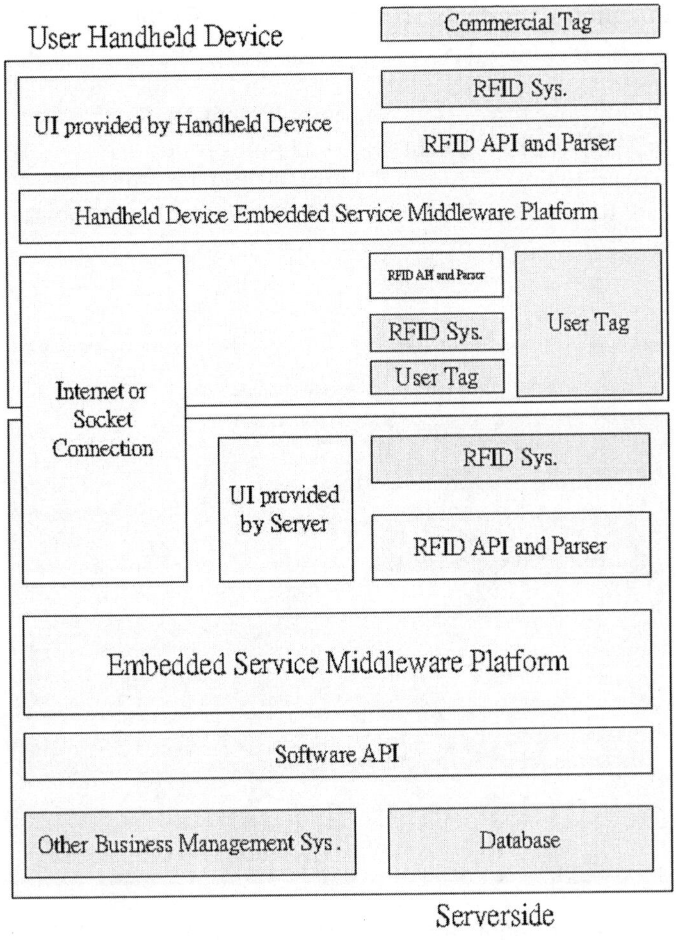

Figure 3. The whole framework of location-aware RFID application service system

2.3. Rfid Applications On Supply Chain Management

2.3.1. Application

Existing RFID applications on supply chain management (SCM) can record something about materials, goods, and products during production [38, 39]. An integrate system with RFID and SCM also can supplies new value-added services such as products secure protection and to query products record [40]. And integrating promising information technologies such as RFID technology, mobile devices-PDA and web portals can help improve the effectiveness and convenience of information flow in construction supply chain control systems [41]. In addition, RFID can be use in a lifecycle of a product to reduce the time which spend to find a product. Therefore, RFID is a technology to reduce the time to identified objects that can improve automation in the traceability management of Supply Chain.

Figure 4. The environment of test and verification

In Warehouse management, many companies have used RFID technology replaced the Bar Code or QR Code as recognition of the key features. Because of the Bar Code and QR Code are limited in existing format companies decided to select a solution improving automation in warehouse management. Then the RFID technology is an efficient technique to solve that problem. Company integrated RFID technology with warehouse management that is not only an electronic process solution but can provided customers with new services such as location information of products, search stock of products, and provide inventory information [42, 43].

In healthcare, RFID also can use to trace patients, blood sampling, drop management, etc. Kumiko Ohashi, SakikoOta, LucilaOhno-Machado, and HiroshiTanaka [44] developed a smart medical environment with RFID technology. This research used two types of frequency on tracking system for tracking clinical intervention such as drug administrations and blood tests at the patient bedside. Furthermore, Chung-Chih Lin, Ping-Yeh Lin, Po-Kuan Lu, and Guan-Yu Hsieh [45] proposed a healthcare integration system for disease assessment and safety monitoring of dementia patients. The proposed healthcare integration system provides the development of an indoor and outdoor active safety monitoring mechanism.

Hence, due to that the RFID technology could provide some services with auto-identify such as administration of drops and samplings, safe monitoring of patients, process control in medical. These new type services can reduce the search time in administration of drops and samplings and human error in medical process. The major advantage of using RFID technology in medical is reduced the human error.

However, RFID applications on Supply Chain Management and Warehouse Management were provided static information as previously noted which helped to deal with problems after accidents. Information on RFID systems was lacking warning data of preventing accidents [5]. To summarize, both Supply Chain Management and Warehouse Management are increase economic values of product. Those operations of management can be help to support product safety information and attribution of responsibility information for customer and enterprise. When enterprise want reduce possible impaired factors to improve the value of product, the first thing should to do is disease management. In addition, disease management has two important methods which to find out the initial pathogen and reduce the spread of pathogens and the infection rate. The spread of pathogens and the infection rate decide the effect areas and damages. Because of enterprise can economic damage control by detect symptoms of product at early stages that will help to reduce cost of operation by itself.

For example, to improve the efficiency of management in cultivation, an *RFID Based Fuzzy Inference Algorithm for Disease Warning and Tracking via Cloud Platform* is proposed. Users could manage the cultivation history, related bio-information, and possible disease tracking. The proposed system modifies the traditional cultivation management system by several fields: 1) first, to shorten the processing time of object recognition in production operation; 2) second, to establish electronic records of production in production management systems; 3) third, to integrate supply chain managements with a central server and provide real-time environment monitoring and plant disease management services for users; and 4) last, to establish an information platform to share with users.

Due to that the contents storage in the memory of RFID tag can be changed when users need. Furthermore, RFID can apply in recognition and also can work in hostile environment such as wet and dirty [26]. RFID provides large read range (or induction range) than Barcode and QR code. Therefore, the RFID system can help to efficiently identify object which equips RFID tag even in non-uniform position. Besides, RFID tag is rewritable. User can remove or rewrite the content of RFID tag when the induction happened between the RFID tag and the RFID reader. In order to overcome the environment factors of cultivation and the objects size, RFID is the solution that can suitable to solve these problems.

In recent years, transportation becomes faster with long distance and also causes more areas infected by disease more easily. After infectious disease influencing a mounts of areas, disease management is more complex and ineffectual [6]. Therefore, quick disease control and prediction is important since it could help to reduce the cost and complex of disease management. Hence, effective disease data tracking and collection of pathogens is necessary. Due to portable RFID tags and non-touched transmission, local area wireless application about disease management for tracking and collection data based on RFID system is proposed. In addition, an information platform which collects data from everywhere and stores the data in its database for the members of supply chain is needed and important. Every platform user can query and access some information from an information sharing platform via network. An information platform can store a lot professional data of a particular field. The platform also can integrate information from each region and has more powerful computing for more information services.

2.3.2. Method

The proposed application system structure is shown as Figure 4. The system infrastructure includes RFID system such as RFID tag, RFID reader, mobile RFID device, etc., and software framework such as database, *Environmental Affection Evaluation Method*, and *Disease Tracking Service* in cloud. The user application layer indicates the mobile RFID device which is used for inducting the RFID tag of local objects such as crops or livestock. The information read from the RFID tag by the RFID reader will be transmitted to the corresponding application and database in cloud. Considering the real implementation, *Environmental Affection Evaluation Method* and *Disease Tracking Service* can be established as the middleware of the whole system or the corresponding applications in cloud. When a user wants to query the information or obtain the disease warning, the proposed *Environmental Affection Evaluation Method* and *Disease Tracking Service* can notify the client user via network from cloud platform.

Generally, an RFID system includes RFID Tags, RFID Readers and Application programs. An RFID tag is a digital storage device that used for identification and information recording. A Reader can access, read, or write data into RFID tags through electromagnetic induction. A user can only use the RFID tag without power consumption. In addition, the mobile device used in RFID system can be a Person Digital Assistant (PDA), a Person Computer (PC) or a laptop (Notebook), which executes the reading and writing actions via RFID systems (include software API and hardware). The middleware mainly manage and deal with the RFID event such as the RFID information sent to or from other systems. After receiving the message from RFID readers, the content of RFID tag can be identified. Then the RFID information will be transmitted and recorded in the database in cloud by *RFID Event Processing*. In addition, the corresponding information sent by *RFID Event Processing* is also presented by the user interface. The *RFID Event Processing* also properly manages and provides the information service for *Environmental Affection Evaluation Method* which analyzes and evaluates the affection degree of the environmental factors.

After inducting the RFID, the information can be transmitted to the corresponding applications and recorded into database in cloud. Each RFID tag will establish an individual object history about the location, resident time, environment state of the RFID tag when it was storaged, etc. To trace and track the potential diseased objects, the *Disease Tracking Service* for client users is needed.

Via *Disease Tracking Service*, customers or client users can obtain active and passive warning message. First, when a customer or client user uses the mobile RFID system to query the information, the corresponding record in the past will be checked and sent back from the database in cloud platform. Then, the members of the supply chain can monitor and manage the state of objects. Second, if the specific object (crop or livestock) is found to be diseased, this object will be marked as the dangerous object. Then, according to the object history, the route, location, etc., where this dangerous object ever passed will be traced back. According to the structure presented in Figure 2, since the object history is established in cloud, the members of the supply chain can exchange the information via cloud. Therefore, the history of the dangerous object can indicate the information about place, location, resident time, etc. Therefore, if the *Disease Tracking Service* finds the dangerous object, the corresponding history, members of supply chain, and the potential diseased that evaluated by*Environmental Affection Evaluation Method* can be notified and traced. In other words, no matter where the objects are, the *Environmental Affection Evaluation Method* can always give the probability value of objects which indicate the potential diseased probability. By using the *Disease Tracking Service*, the location, warehouse, manager, etc., will be notified that how many objects with the different and individual potential diseased probability currently reside at or ever passed the place. Therefore, the object with high potential diseased probability can be discovered in early phase. Figure 5. presents the implementation of the system.

The tracing and tracking system based on RFID system not only provide original services in supply chain or warehouse which includes the records of products for managing and querying, but also provide new services such as to obtain the potential probability of diseases and send the warning sign for users. Furthermore, the system can derive the possible disease infecting area for users to control and update the latest information of production for users to track and trace. In addition, the verification shows that the proposed system is realistic and can provide the public and personal services automatically. By using this innovation RFID system, users could get most information to prevent disease in agricultures that helps users to reduce the cost of production, control the range of disease occurrence, and providing a warning for disease prediction.

3. DISCUSSION

Radio Frequency identification (RFID) is the popular wireless induction system [7] [2, 3, 8, 30]. Each RFID tag in an RFID system is equips a unique ID (UID) itself. UID can help to shorten the identification time for individual object recognition. In general, there are several methods to achieve aim of Automatic Identification such as Barcode and QR code. However, due to that the Barcode and QR code have the limitation in environmental affection such as wet or water, to maintain the usability and the reliability of the Barcode or QR code is too difficult. Therefore, using RFID can be the solution which provides the distance induction with better characteristics such as anti-water and rewritable memory.

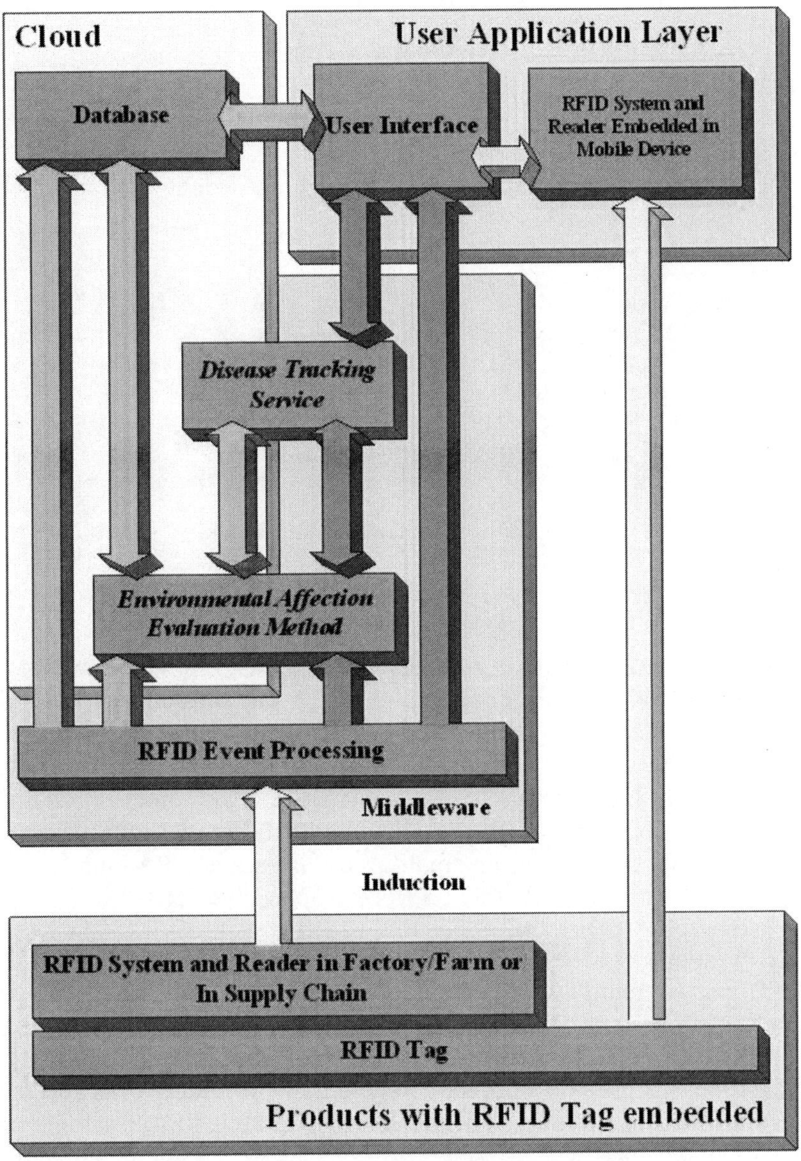

Figure 5. The structure of RFID system with Cloud environment

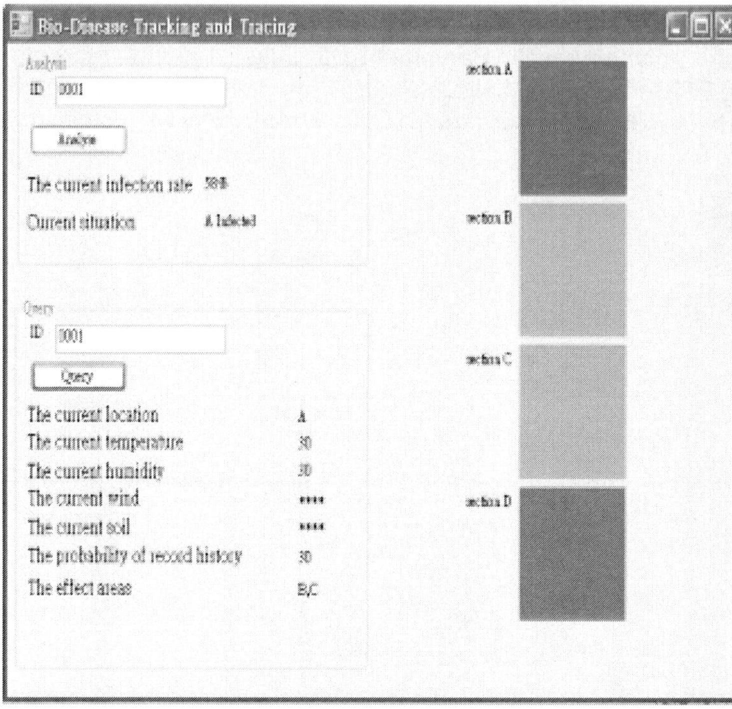

Figure 6. The implementation of tracing and tracking system based on RFID

A standard RFID system consists of Tag, Reader, Middleware, and Application. When an independent RFID tag approaches the RFID antenna, the induction between RFID tag and antenna happens [9]. The RFID antenna reads or obtains the information and content recorded in the tag. Then the information is translated into the computational data by the RFID reader. Due to the portable RFID tag and untouched data transmission, many local or small area wireless applications for track and trace based on RFID systems were proposed [2, 7, 8].

RFID reader can access data of RFID tag and transmit the content from RFID tag to middleware which is a necessary component in RFID application system. The middleware is also the interface software that connects new RFID hardware with legacy enterprise IT systems [36]. Middleware is used to route data between the RFID networks and the IT systems within an organization. It merges new RFID systems with legacy IT systems.

RFID Reader is also called Interrogator. The RFID reader can read and write data of RFID tag via radio frequency. RFID readers can classify serial reader and network reader according to connection interface.

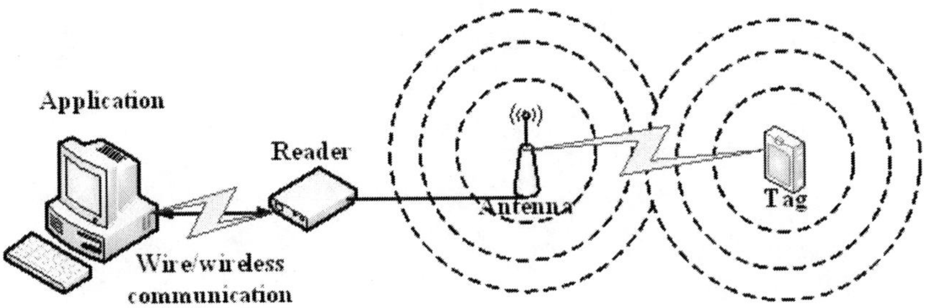

Figure 7. Operation concept of RFID system

In Radio Frequency Identification, there are four standard frequency ranges used: Low Frequency (LF), High Frequency (HF), Ultra-HF (UHF), and Microwave respectively. Frequency decides the reading distance of RFID devices and the interfered with environment. Figure 7 shows the shows the conditions and factors related to the frequency. Higher frequency RFID tag has longer induction distance, higher data rate, and smaller Tag size. On the contrary, lower frequency RFID tag has shorter induction distance, lower data rate, and bigger tag size. In addition, higher frequency RFID tag has bad performance when tag near metal or liquids. In this thesis, the proposed system selects low frequency RFID tag because that the cultivation environment is wet and dirty. Low frequency RFID tag has better performance than high frequency RFID Tag at cultivation environment.

There are two main types of RFID tag, active and passive. Active tag is a powered tag which equips battery and actively sends signal itself. In opposition to active tag, passive tag without battery can only send signal when the tag is inducted. Table 1 presents an introduction for RFID Tag.

The RFID technology is affected by several factors such as frequency, energy, and environment. When users decide to use RFID in a particular place, users have to select the fittest specification of RFID at first. Because the fittest specification of RFID can make higher performance in the efficient of RFID application system, to select the fittest specification of RFID is important. Furthermore, the cost of RFID is an important factor for users to select.

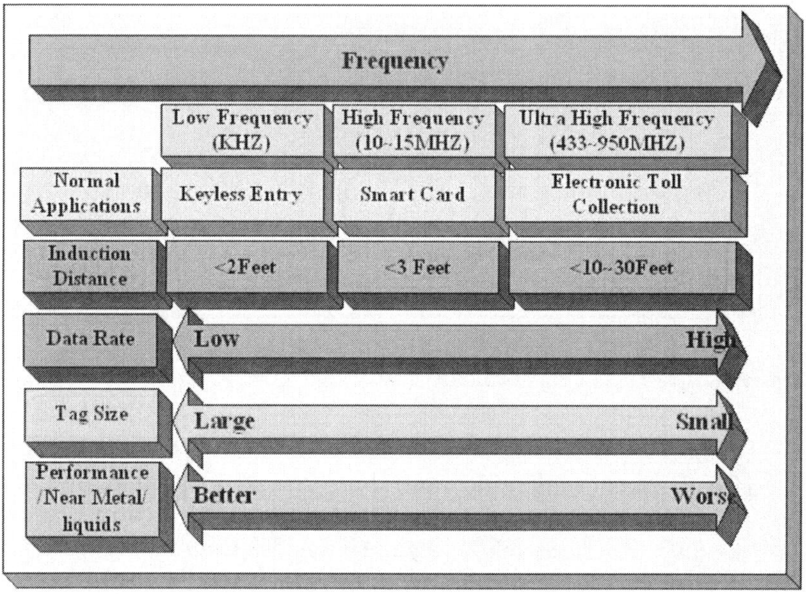

Figure 8. RFID frequency specification chart

4. CHALLENGES

4.1. Cost

Today, based on the behavior of customers, the history or record of the production including the delivering is not the most important thing. Most consumers care about the price, expiration date, or packing of the goods. In other words, although the RFID can enhance the management of logistic, to embed the RFID technology into current system also cost a lot. For example, one RFID tag for paste once (non-reusable) may cost \$0.8. However, a bottle of water may also cost about \$1. Most consumers would not to pay almost twice payment to obtain the information which they do not care about. Comparing with barcode or 417-barcode, the RFID tag costs more than barcode or 417-barcode which can be printed by a printer. In other words, to identify an object via RFID tag costs more than using barcode.

Table 1. The relational table of RFID frequency and instruction.

Freq. Effect	**Low Frequency(LF)**	**High Frequen cy (HF)**	**Ultra-HF (UHF)**	**Microw ave**
Usable Frequen cy	100~500 KHz	10~15 MHz	433~95 5 MHz	1GHz~
Commo n Frequen cy	125 KHz 135 KHz	13.56 MHz	433 MHz 868~95 0 MHz	2.45 GHz 5.8 GHz
Power Type	Passive	Active / Passive	Active / Passive	Active / Passive
Reading Distance	Short range	Short range	Longer range	Longer range

[i] - Source: EPCglobal.(http://www.gs1tw.org/twct/web/EPC/ index. jsp)

In opposition to once-using RFID tag, to reuse the RFID tag may be a solution. A plastic card where RFID embedded can be used for identification or payment. For example, in Taiwan, most mass transit systems can be paid according to the RFID card that pre-registered and sold to the consumers. Each consumer can use the RFID card to pay the consumption in convenient store, fee of the parking lot, and use the card as a ticket for railway or MRT system. Due to the convenience, most consumers have at least one RFID card for payment. However, the security for financial application is very important. Although there are some security method proposed, no algorithm or system can provide the 100% promise that the security method is always safe.

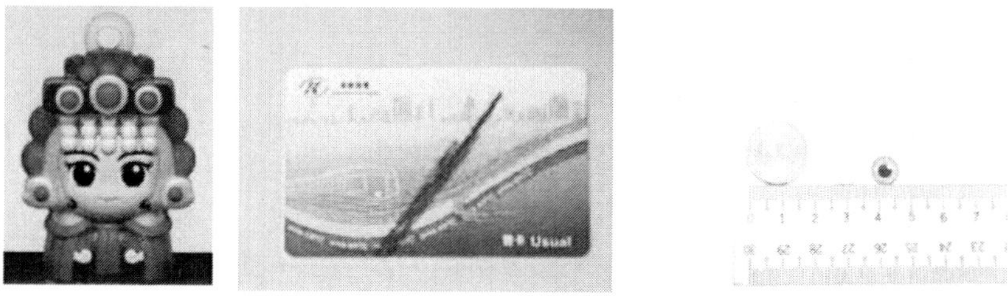

(a) (b) (c)

Figure 9. The sample of RFID Tag; a.3D Toy, b. plastic card, c. button tag.

In addition to the RFID card, various styles of RFID tag may be required by different consumer. According to the environment, the RFID tag may be designed to anti-water or anti-wet. Some RFID tags need to be applied to the metal environment such as containers. Due to the characteristics of RFID, suitable frequency of RFID corresponding to the environment should be selected carefully. In other words, the cost of different RFID may be also different according to the applications and environment. Furthermore, different shapes of RFID tag also costs. For example, the cost of an RFID tag embedded into a 3D toy is much more expensive than that embedded in a plastic card.

4.2. Hardware and Integration

The RFID Antenna is the main component for RFID tag induction. The antenna continuously spreads the electromagnetic wave. The energy is transmitted to the RFID tag. After induction, the RFID Antenna also receives the signal from the RFID tag.

After receiving the signal, the RFID Reader translates the signal into the digital data such as the UID of this RFID tag. Then, the RFID Reader sends the digital data to the corresponding systems or applications. To implement the RFID system, not only the RFID tag but also RFID reader and antenna hardware should be considered. Due to the design and product limitation, the RFID antenna cannot dynamically change.

Therefore, similar to the RFID tag, the environment of the RFID system affects the size and cost of the antenna. The size of the antenna increases, the costs also increases. In addition, considering the implementation environment, to place the antenna at the suitable location for signal receiving also affects the performance and costs. Therefore, to integrate the RFID system with the existed system, some additional problems may need to be overcome.

The existed applications or systems should include or integrate the RFID system. In other words, to integrate the RFID system, the original working practices of the existed system may be changed which needs extra costs. For example, to identify goods in warehouse, original applications or systems may only need the manual operation. However, to integrate the RFID system, some infrastructure such as the placement of RFID antenna, wire line for connection between antenna and reader, and the establishment of RFID reader and system server are required. In other words, the extra costs of RFID infrastructure are needed. In addition, some existed systems are based on mechanical operation without too much intelligent analysis. For example, a car parking lot only needs to open the gate when a car approaching or according to the teleswitch. When integrated with the RFID system, the RFID antenna should be placed in front of the gate for induction. All the cars to the parking lot should present the RFID tag given on demand. In addition, the RFID reader should be used to analyze the signal information from the RFID tag. The server which includes the database should be used to judge whether the gate should open or not. Although the RFID system enhances the automation with less manual operation, some extra costs and delay may also happen. Therefore, the benefits of RFID system integration such as automation, information exchanging with third party applications, etc., are very important. Only when the benefits or additional new functions overcome the extra costs of RFID system integration then the integration of system will be used.

4.3. Plug and Play Middleware

In the RFID systems or applications, there are two partitions: RFID devices (includes RFID tag, antenna, and reader) and other devices or systems. Therefore, the application or middleware for communicating these two parts is needed. When using the RFID device, the third party systems or applications should obtain the information from RFID devices.

Due to that there are many types of RFID hardware, the application program interface (API) for the communication between RFID Hardware and different third party applications is needed. In addition, the end user's devices are also various. Hence, the plug and play middleware for different hardware and applications is important.

To manage the RFID information from different RFID Hardware, and the communication with different applications, the Plug and Play Middleware is proposed. To realize the concept of Plug and Play, the proposed middleware has to manage the information from the all possible third party RFID Hardware, deal with and parse the information, and then provide the required information to the corresponding applications. Therefore, the main purposes of the proposed Plug and Play Middleware are:

1) to parse the information from the RFID Hardware. Due to that there are different RFID product, the RFID parser is needed for analyzing and parsing the information from RFID Hardware. The information about UID, password, etc. will be parsed as the string for the further execution of applications. In this paper, two possible parsers are established. First, the Plug and Play Middleware provides the remote procedure call (RPC) function for the third party RFID Hardware. The UID of the RFID tag inducted by the RFID Hardware will be formulated as the string. In addition, the password or requirements for further information such as decryption code recorded in the End User RFID Device can be provided by the remote procedure call function. Second, for general communication, the Plug and Play Middleware also provides the sever-client socket link between the RFID Hardware and the middleware. In other words, even the RFID Hardware cannot implement the remote procedure call, depends on sever-client socket link, the information can be transmitted between Plug and Play Middleware and RFID Hardware.

2) to provide the application program interface (API). Since the RFID Hardware may not directly communicate with the applications, the Plug and Play Middleware has to implement the corresponding API for other third party applications or software.

Furthermore, the Plug and Play Middleware also should implement two possible APIs: the external procedure call and network communication. If the application is embedded in the Plug and Play Middleware, the external procedure call sends the required information to the specific application. In addition, some communications of the related applications such as database query are also established by the external procedure call. Then, the Plug and Play Middleware deals with the results from the external procedure call. In opposition to external procedure call, for the concept of Plug and Play, normal network communication should also be implemented. Most third party software or applications can communicate with the Plug and Play Middleware via sending the information in string format. For example, if the third party application requires the further checking, the Plug and Play Middleware sends the required information such as UID to the server via Internet. To reduce the cost for communicating with different third party applications, the unify data storage format is necessary. Therefore, the eXtensible Markup Language (XML) can used as a data exchange standard. After obtaining the response from the server, the Plug and Play Middleware can acknowledge the third party application. At last, the corresponding services can be presented.

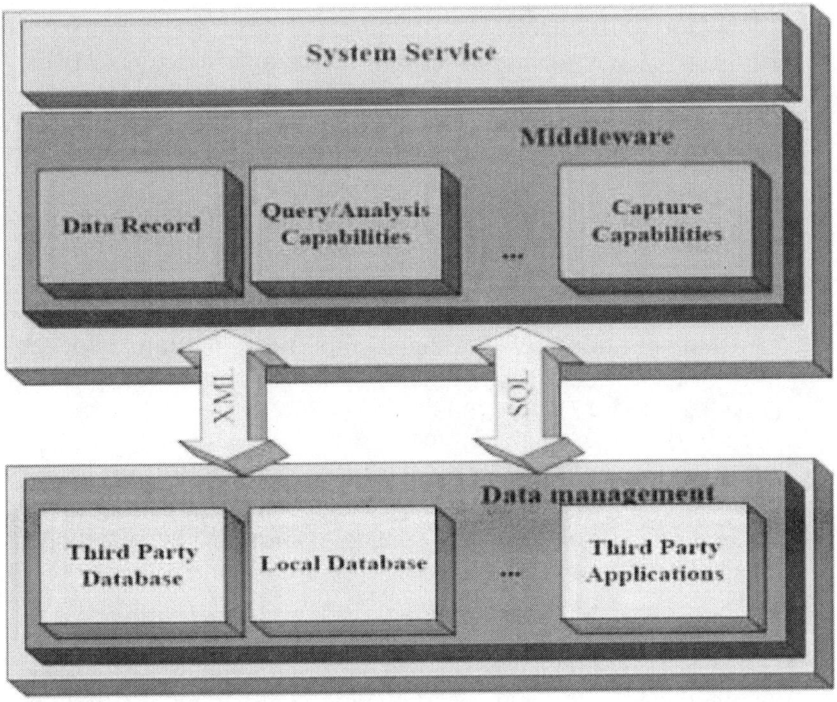

Figure 10. The structure chart of data exchange in Database management

5. CONCLUSION

In this chapter, we show applications and systems based on RFID technology which integrated into the existed service systems. The RFID technology can enhance the automatic management procedure. Identification and tiny information exchanging can be achieved. Individual or personal services can be provided to different consumers. However, to establish the RFID embedded systems and applications, the cost, convenience, feasibility should be considered. To adopt RFID system, some extra costs such as RFID tag and hardware should be overcome by the enhanced performance of management. In other words, to implement the RFID systems for the consumers, to enhance the convenience for consumers will be an important issue than the cost.

REFERENCES

1. "Hoboken RFID-enables Its Parking Permits," RFID Journal, Jun 2006, http://www.rfidjournal.com/article/articleview/2421/1/1/

2. "RFID Takes a Swing at Ticket Fraud", RFID Journal, Dec 2005, http://www.rfidjournal.com/article/articleview/2060/1/1/

3. "Moscow Metro Tries RFID-Enabled Ticketing," RFID Journal, Feb 2007, http://www.rfidjournal.com/article/view/3049/

4. A.S. Martinez-Sala, E. Egea-Lopez, F. Garcia-Sanchez, J. Garcia-Haro, "Tracking of Returnable Packaging and Transport Units with active RFID in the grocery supply chain," Computers in Industry, Vol. 60, Issue 3, pp.161-171, Apr 2009.

5. T.C. Poon, K.L. Choy, Harry K.H. Chow, Henry C.W. Lau, Felix T.S. Chan, K.C. Ho, "A RFID case-based logistics resource management system for managing order-picking operations in warehouses," Expert Systems with Application, Vol. 36, Issue 4, pp. 8277-8301, 2009.

6. C. Loyce, J.M. Meynard, C. Bouchard, B. Rolland, P. Lonnet, P. Bataillon, M.H. Bernicot, M. Bonnefoy, X. Charrier, B. Debote, T. Demarquet, B. Duperrier, I. Felix, D. Heddadj, O. Leblanc, M. Leleu, P. Mangin, M. Meausoone, G. Doussinault, "Interaction between cultivar and crop management effects on winter wheat diseases,"

lodging, and yield, Crop Protection, Vol. 27, Issue 7, pp. 1131-1142, 2008.

7. Colin H. Burton, "Reconciling the new demands for food protection with environmental needs in the management of livestock wastes," OECD Workshop, Vol. 100, Issue 22, pp. 5399-5405, 2009.

8. Beijing Olympic Games Prompts RFID Development in China, http://www.rfidglobal.org/news/2007_9/200709031653253861.html

9. M.S. Jian, K. S. Yang, and C.L. Lee, "Context and Location Aware Public/Personal Information Service based on RFID System Integration," WSEAS Trans. on Systems, Vol. 7, pp. 774-784, Jun. 2008.

10. Z. Pala and N. Inanc, "Smart Parking Applications Using RFID Technology," Proc. of 1st Annual RFID Eurasia, pp. 1-3, Sep 2007.

11. M.F. Lu, S.Y. Chang, C.M. Ni, J.S. Deng, and C.Y. Chung, "Low Frequency Passive RFID Transponder with Non-revivable Privacy Protection Circuit," Proc. of WSEAS Inter. Conf. on Instrumentation, Measurement, Circuits, and Sys., pp. 166-169, April 2006.

12. "Hospital Uses RFID for Surgical Patients," RFID Journal, Jul 2005, http://www.rfidjournal.com/article/articleview/1714/1/1/

13. "RFID Hospital: Columbus Children's Hospital To Install RFID System From Mobile Aspects", RFID Solution Online, Mar 2007.

14. "RFID trial tracks hospital equipment," http://www.computing.co.uk/computing/news/2168717/rfid-trial-tracks-hospital

15. C.L. Lai, S.W. Chien, S.C. Chen, and K. Fang "Enhancing Medication Safety and Reduce Adverse Drug Events on Inpatient Medication Administration using RFID," WSEAS Trans. on Communications, Vol. 7, pp. 1045-1054, Oct 2008.

16. M.S. Jian and S.H. Hsu, "Location Aware Public/Personal Diversity of Information Services based on embedded RFID Platform," Proc. of ICACT'09, pp. 1145-1150, Feb 2009.

17. Ming-Shen Jian, Kuen Shiuh Yang, and Chung-Lun Lee, "Modular RFID Parking Management System based on Existed Gate System

Integration," WSEAS Trans. on Systems, Vol. 7, pp. 706-716, Jun 2008.

18. http://www.ist.com.tw/

19. C.S. Baptista, C.P. Nunes, A.G. de Sousa, E.R. da Silva, F.L. Leite, and A.C. de Paiva., "On Performance Evaluation of Web GIS Applications," Proc. of the IEEE 16th Inter. Workshop on Database and Expert Sys. App., pp. 497-501, 2005.

20. M. Marmasse and C. Schmandt, "Locationaware information delivery with commotion," Proc. of 2nd Inter. Symposium in Handheld and Ubiq. Comp., 2000.

21. D. Ashbrook and T. Starner, "Learning significant locations and predicting user movement with GPS," Proc. of Inter. Symposium on Wearable Comp., 2002.

22. D.J. Patterson, L. Liao, D. Fox, and H. Kautz, "Inferring High-Level Behavior from Low-Level Sensors," Proc. of 5th Inter. Conf. on Ubiqutious Comp., 2003

23. P. Bahl and V.N. Padmanabhan, "RADAR: An inbuilding RF-based user location and tracking system," Proc. of Infocom, pp.775-784, 2000.

24. [24]K. Romer, T. Schoch, F. Mattern, and T. Dubendorfer, "Smart Identification Frameworks for Ubiquitous Computing Applications," Wireless Networks, pp. 689-700, 2004.

25. S. Willis and S. Helal, "A Passive RFID Information Grid for Location and Proximity Sensing for the Blind User," TR04-009, 2004.

26. G. Chen and D. Kotz, "A Survey of Context-Aware Mobile Computing Research," Dartmouth College Technical Report TR2000-381, 2000.

27. J. E. Bardram, R. E. Kjar, and M. O. Pedersen, "Context-Aware User Authentication - Supporting Proximity-based Login in Pervasive Computing," Proc. of Ubicomp, pp. 107-123, 2003.

28. C. Floerkemeier and M. Lampe, "Issues with RFID usage in ubiquitous computing applications," Proc. of Pervasive Comp., pp. 188-193, 2004.

29. M. Crawford, et. al., "RFID Enabled Awareness of Participant's Context in eMeetings," Proc. of Pervasive Tech. Applied Real-World Experiences with RFID and Sensor Net., 2006.

30. J. Bravo, R. Hervas, G. Chavira, and S. Nava, "Modeling Contexts by RFID-Sensor Fusion," Proc. of Pervasive Comp. and Comm. Workshops, pp. 30-34, 2006.

31. J. Choi and E. Kim, "Using of the Context in RFID Systems" Comm. of the Korea Information Science Society, pp. 64-70, 2006.

32. L. Han, S. Jyri, J. Ma, K. Yu, "Research on Context-Aware Mobile Computing," Proc. of 22nd Inter.Conf. on Adv. Infor. Net.g and App. – Workshops, pp. 24-31, 2008.

33. L. Buriano, "Exploiting Social Context Information in Context-Aware Mobile Tourism Guides," Proc. of Mobile Guide 2006, 2006.

34. M. A. Munoz, M. Rodriguez, J. F. Center, A. I. Martinez-Garcia and V. M. Gonzalez. "Context-Aware Mobile Communication in Hospitals," IEEE Computer, pp. 38-46, 2003.

35. C. Ciavarella and F. Paterno, F., "The design of a handheld, location-aware guide for in-door environments," Springer Verlag Personal and Ubiquitous Computing, pp. 82-91, 2004.

36. Z. Pala and N. Inanc, "Smart Parking Applications Using RFID Technology," Proc. of 1st Annual RFID Eurasia, pp. 1-3, Sep 2007.

37. R. Tesoriero, J. A. Gallud, M. Lozano, and V. M. R. Penichet, "A Location-aware System using RFID and Mobile Devices for Art Museums," Proc. of 4th Inter. Conf. on Autonomic and Autonomous Sys., pp. 76-82, 2008.

38. Y.C. Hsu, A.P. Chen, C.H. Wang, "A RFID-Enabled Traceability System for the Supply Chain of Live Fish," Automation and Logistics, pp. 81-86, 2008

39. P. Jones, "Networked RFID for use in the Food Chain," Proc of the 2006 Emerging Technologies and Factory Automation Conference, pp. 1119-1124, 2006.

40. F. Gandino, B. Montrucchio, M. Rebaudengo, and E.R. Sanchez, "On Improving Automation by Integrating RFID in the Traceability

Management of the Agri-Food Sector," IEEE Transactions on Industrial Electronics, Vol. 56, Issue 7, pp. 2357-2365, 2009.

41. L.C. Wang, Y.C. Lin, and P.H. Lin, "Dynamic mobile RFID-based supply chain control and management system in construction," Advanced Engineering Informatics, Vol. 21, Issue 4, pp. 377-390, 2007.

42. A.S. Martinez-Salaa, E. Egea-Lopez, F. Garcia-Sancheza and J. Garcia-Haroa, "Tracking of Returnable Packaging and Transport Units with active RFID in the grocery supply chain," Computers in Industry, Vol. 60, Issue 3, pp. 161-171, 2009.

43. M. Tu, J.H. Lin, R.S. Chen, K.Y. Chen, and J.S. Jwo, "Agent-Based Control Framework for Mass Customization Manufacturing With UHF RFID Technology," IEEE System Journal, Vol. 3, Issue 3, pp.343-359, 2009.

44. K. Ohashi , S. Ota, L. Ohno-Machado, and H. Tanaka, "Comparison of RFID Systems for Tracking Clinical Interventions at the Bedside," Proc of AMIA Annu Symp 2008, pp. 525–529, 2008.

45. Chung-Chih Lin , Ping-Yeh Lin, Po-Kuan Lu, Kuan-Yu Hsieh, Wei-Lun Lee, and Ren-Guey Lee, "A Healthcare Integration System for Disease Assessment and Safety Monitoring of Dementia Patient," IEEE Transactions on Information Technology in Biomedicine, Vol. 12, No. 5, pp. 579-586, 2008.

2

A SYSTEMATIC REVIEW OF RFID APPLICATIONS AND DIFFUSION: KEY AREAS AND PUBLIC POLICY ISSUES

Kwangho Jung Sabinne Lee

Korea Institute of Public Affairs, Graduate School of Public Administration of Seoul National University

ABSTRACT

RFID applicants called as e-ID, smart tag, and contactless smart card are being applied to numerous areas in our daily life, including tracking manufactured goods, currency, and patients to payments systems. To review these various applications of RFID is important to exploring not only ongoing e-governance issues such as digital identification, delivery process, and governance but also business oriented application areas like supply chain. Through a systematic review methodology from 111 previous studies about RFID technology for public sector, we found six key areas of RFID applications: defense and security, identification, environmental applications, transportation, healthcare and welfare, and agriculture-livestock. We also suggest that the diffusion and applications of RFID can involve unexpected disadvantages including technological deficiency, uncertain benefits, dubious transparency, uncomfortable privacy issue, and unequal distribution of digital power and literacy. Further research on RFID impact includes not only various theoretical issues of but also legal and managerial problems. Rigorous research is required to explore what factors are critical to adopt and implement new RFID applications in terms of technology governance and digital literacy. Massive data driven research is also expected to identify RFID performance in government agencies and various industry sectors.

KEYWORDS

RFID, Digital identification, Digital delivery, Smart tag, Contactless smart card

1. BACKGROUND

RFID technology has been widely implemented all over the world and its impact on our daily life is very diverse and massive (Li et al., 2006; Wyld, 2005). Those diverse areas of RFID application include logistical tracking, monitoring and maintenance of products, product safety and information, and payment process. Today many governments around the world in both developed[1]and developing[2] countries are trying to apply it for various areas from tracking manufactured goods, currency, and patients to securing sagety of payments systems. Massive RFID applications around all the industry sectors and countries are expected to generate a huge potential benefits for sustainable efficient energy infrastructure, transportation safety, and health care. Over the past 50 years, RFID technology went through innovations and progressions to become a more efficient and effective gadget for human beings as well as effective solutions of technical and organizational problems in various industry sectors. However, key issues of appropriate ICT technology, governing networks among RFID domains, standardization requirement, and privacy still remain unsolved[3].

We review previous literature about RFID technology used in public sectors in order to identify what has been done and found to suggest policy implications and further research agenda. More specifically, we discuss four aspects regarding RFID research issues and policy implications. First, we examine various competing concepts of RFID use by governments all over the world. Second, we categorize numerous applications of RFID technology through analyzing previous literature. Third, we try to Figure out technological issues and governance problems that RFID technology faces today. Last, we draw key public issues and suggest future research agenda.

2. METHODOLOGY OF THE RFID LITERATURE REVIEW

2.1 A brief history of RFID technology

RFID technology was emerged as Frederick Hertz found existence of radio frequency during his experiment in 1886 (Wyld, 2005) and developed for the purpose of defense during the Second World War[4]. During 1970s and 1980s, the RFID system attracted plenty of scholars and innovators, so efforts to register patents progressed (Takahashi, 2004). Researchers like Charles Walton had registered a patent to use RFID. In the 1980s, many US and European companies recognized the importance of developing RFID technology and started to manufacture RFID tags. Soon scholars at MIT University opened an Auto-ID center to promote the use and implementation of RFID technology. But most of the scholars report that the first commercialization of RFID technology was done by Wal-Mart as they launched RFID based material identifying system in 2005 (Shahram and Manish 2005). Wal-Mart is now tracking merchandise including food, apparels, and electronic items with RFID technology in their supply chain.[5]. RFID technology is a brand new policy tool that can ensure high transparency, efficiency and effectiveness not only in industrial areas but also in government service delivery. Table 1 describes a brief history of how RFID technology was developed and diffused.

2.2 Research design for a systematic review

We searched online data base and expert based information to identify RFID publications between 2003 and 2015. We categorized RFID applications and analyzed issues and concerns that RFID faces today by systematically reviewing published literature. We have collected literature we use for systematic review from two different resources. First, most of the studies are found by searching the e-database. We could access electronic databases, such as Google Scholar, World Web of Science (WWS), Proquest Central, and Science Direct through Seoul National University's main library homepage. We had set 'RFID technology', 'RFID government,' 'RFID application, and 'RFID issue' as keywords for searching literature. We found most of the research through this method of searching. The second method we used for collecting data was having discussions with experts. To do this, we first made a list of experts who specialize in IT technology, Science technology, and public administration. Five experts agreed to help us and recommended some research papers that were known for their fluent flow of logic and plentiful contents. We chose relevant research papers from among experts' recommendations. In sum we had used previous literature collected from two methods we discussed above, searching e-database and asking experts, as our resource of searching.

Table 1 A brief history of RFID technology

Date	Event
1886	The idea of using Radio Frequency to reflect waves from objects was started from Frederick Hertz's experiment.
1930–1940	American navy research laboratories developed a system known as IFF (Identify Friend or Foe).
1940–1950	The first application of RFID consisted of identifying allied or enemy planes during WW2 through the use of IFF system.
1973	Charles Walton, a former IBM researcher registered patent using RFID technology, a radio-operated door lock.
1980–1990	Many US and European companies started to manufacture RFID tags.
2003	The Auto-ID center for MIT became EPC global, an organization whose objective is to promote the use and adoption of RFID technology.
2005	Wal-Mart launched and RFID pilot.

[Figure 1] shows analytical frame that we use for this study. We have determined the literature for systematic review according to three stages shown on the flow chart. First, the original total number of studies we have found from the e-database was 4260. Also 185 research papers were found from experts' recommendations and previous public papers. A total of 4,445 studies were chosen through the first stage. Second, we excluded 4,121 following general eligibility criteria by screening title and abstract. More specifically, we excluded RFID studies only with one of the following criteria: 1) studies focusing on private sector; 2) studies without considering how public sector implemented RFID technology; 3) studies that did not discuss any social scientific implications; and 4) studies that only deal with RFID technology from pure scientific and engineering points of view. In sum, we included only 324 papers that discussed RFID issues and their implications in public sector. Third, we removed further 213 studies too much focusing on private sector or RFID technology itself, rather on its applications in our society and social scientific implications. Finally, 111 articles were chosen for our systematic review.

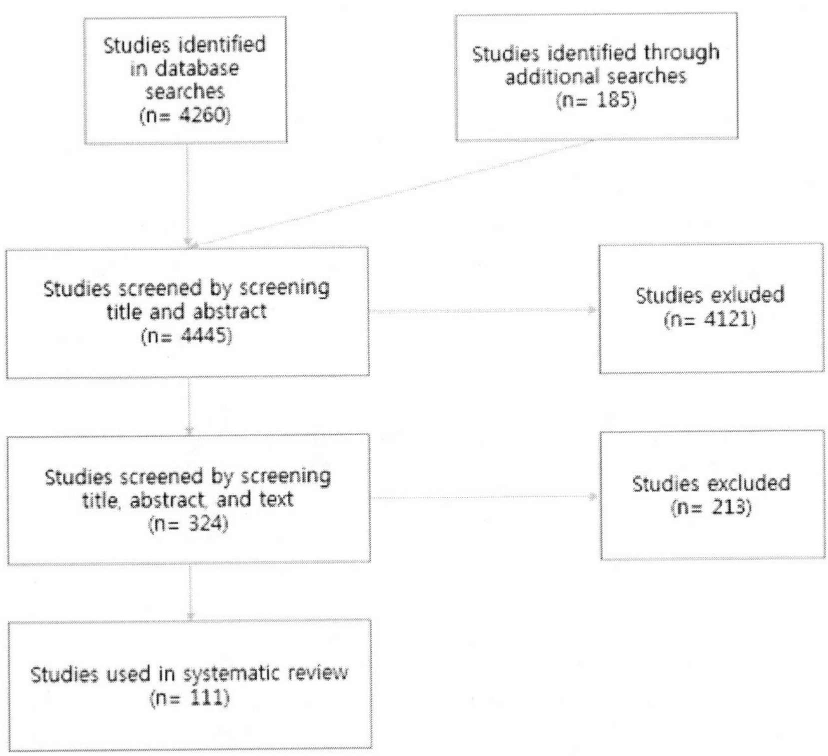

Figure. 1 Analytical Frame

[Figure 2] below showed descriptive statistics of collected literatures by published year. It shows 22 studies were published in 2007 among 111 literatures. As we already described above in history of RFID section, the popularization and commercialization of RFID technology was started in 2005 with Wal-Mart's adoption. It seems that after Wal-Mart's innovative footsteps hit the world, many scholars were started to recognize the potential of new technology and tried to understand and develop RFID technology. Besides some governments from all over the world implemented new way of public service delivery using RFID technology. Consequently, 49 literatures were publishedbetween 2006 and 2008 and it forms almost 45 % of our collected studies

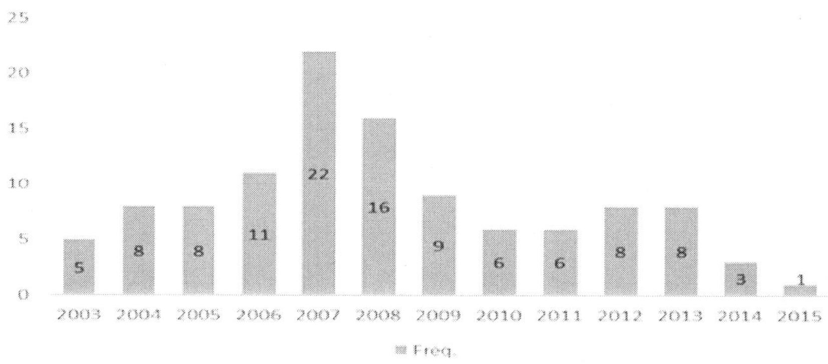

Figure. 2 Descriptive Statistics of Literatures by year

For this study, we categorized governments' way of using RFID technology in 6 areas; Agriculture and Livestock, Defense and Security, Environmental Applications, Healthcare and Welfare, Identification, and Transportation. [Figure 3] shows descriptive statistics of collected literatures categorized by applications. We categorized studies that did not focus on specific sector and analyze and introduce RFID technology from the general perspectives as 'RFID general'. 'RFID general' studies usually deal with various ways of using RFID technology in diverse sectors simultaneously. As we can see from [Fig. 3], RFID general area had 42 papers. That means still lots of RFID studies could not be fully specialized and remained in status of generally introducing RFID technology. Identification sector scores secondly highest number of published literatures among areas. This result seems natural because e-ID card or e-Passport have most powerful force that can hurt privacy, one of the most serious and notorious issues that RFID technology face today

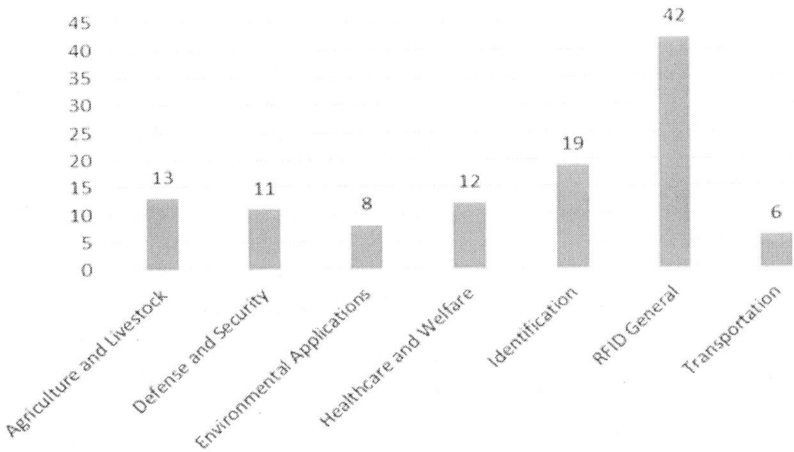

Figure. 3 Descriptive Statistics of Literatures by applications

3. KEY APPLIED AREAS OF RFID

3.1 Defense and security

As we show in [Table 1], the history of RFID technology was started from the need for ensuring national security. Almost 60 years have passed since US army developed RFID based identification system to identify allies and enemies, RFID technology is still used for protecting people. For instance, Weinstein (2005) and Konsysnki & Smith (2003) reported how the US Army and Navy implement RFID technology in cargo containers to identify materials. The US Army and Navy implement RFID not only to identify US troops' own weapons and containers but also to identify enemies in battle (Tien 2004)[6]. RFID systems are also important in terms of airport and port security. After the 9/11 terror attack on the United States, President George W. Bush let all the airport and port in US adopt identification systems based on RFID technology to protect its nation from additional terrorist attacks (Werb and Sereiko 2002)[7]. In 2012, the Taiwanese government decided to implement an RFID based e-Seal system to increase security and efficiency (Tsai and Huang 2012). In addition, RFID technology can be used effectively in prison management[8] and child protection. In some countries like Japan and Republic of Korea, the RFID tag is implemented in child protection monitoring (Table 2).[9]

Table 2 RFID application in Defense and Security

Application	Author	Year	Case	Country
Army and Navy	Weinstein	2005	US Navy embedded RFID into cargo container	US
	Konsynski and Smith	2003	US Army track containers of material	US
	Tien	2004	US Army using RFID for tracking its army in Iraq	US
	Anon	2002	US Army piloted 4 projects using RFID	US
Airport and port Security	Werb and Sereiko	2002	2 RFID programs to prevent terrorist's attack	US
	Smith and Konsynski Zhang	20032013	New York city government project using RFID,*CargoMate*	New York City, US

Application	Author	Year	Case	Country
	Tsai and Huang	2012	Kaohsiung port adopt RFID system for port security	Taiwan
Prison management And Child protection	Kim	2008	Calpatria prison issued RFID embedded bracelet to its inmates	US
	Nicholas	2008	LA County prison management	US
	Ema and Fujigaki	2011	Let parents know the exact time of a child's arrival and departure time	Japan
	Yonhap news	2013	Child monitoring through RFID tag in beach	ROK

3.2 Identification

Electronic passports like 'e-passports' were adopted electrically after the 9/11 attack. After the terrible tragedy broke heart of United States, the American government became aware of the importance of checking VISAs and passports correctly. The US Department of State soon let people who wanted to enter US to use RFID tag embedded electronic passports instead of traditional barcode based passports[10]. The European Union also endorsed the inclusion of biological information in e-passports. The EU Justice and Home Affairs Council decided to include fingerprints as a second mandatory identifier on passports in 2004.[11]In addition, RFID can be used in e-ID cards in various countries. For example, in the United Kingdom, Prime Minister Tony Blair and his Labor Party convinced the nation to adopt biometrically-enhanced national identification cards (Ezovski and Watkins2007). Tony Blair's administration announced its will to implement RFID tag embedded national identification card in late 2004. China is another case where the e-ID card is used today. As a matter of fact, China is the country where e-ID card is widely and largely adopted today. The Beijing Olympics held in 2008 lit the fuse of adoption. The largest smart card project was implemented as a part of preparing the most prominent international sports event. In 2008, the Chinese government supplied 1.2 billion dollars of RFID readers and 2.25 billion dollars of RFID embedded smart cards to citizens. This made China the world's largest market for RFID (Kovavisaruch and Suntharasaj 2007) (Table 3).

Table 3 RFID application in identification

Application	Paper	Year	Case	Country
e-passport	Meingast et al.	2007	US Visit Visa Waiver Program(VWP) focusing on why US government adopted RFID based e-passport	US
	Lorenc	2007	History of e-passport	US
	Ramos et al.	2009	Disadvantage of using e-passport from privacy perspective	US
	Kim	2008	Pros and cons of ROK government using RFID embedded e-	ROK

Application	Paper	Year	Case	Country
			passport	
	Hossain	2014	How Australia adopt and implement e-passport	Australia
e-ID	Ezovski & Watkins	2007	e-ID used in United Kingdom	UK
	Kovavisaruch & Suntahrasaj	2007	Brief history of Chinese e-ID card	China

3.3 Environmental applications

RFID technology can be widely applied in environmental applications. Adopting an RFID system in waste management is the most prominent way of using RFID to ensure efficient, eco-friendly waste management among lots of countries in the world. PAYT (Pay-As-You-Throw) program done by European Union (EU) is the pioneer of this field. PAYT is an RFID based waste pricing model that allows each individuals or each household to pay for the tag along with the total amount of waste they throw. Since each household and individual has a waste box in which RFID tags are embedded, the exact volume of waste can be calculated. In Europe this incentive based system has been proven to be a powerful policy tool for reducing the total amount of waste and for encouraging recycling (Schindler et al. 2012). Similar systems are broadly implemented in US (Ransford et al. 2012). In South Korea, the ministry of environment introduced it to industry and urged them to use an RFID based waste management system, especially in medical waste management. RFID technology is implemented in waste management in developed and developing countries, but the purpose of adoption is somewhat different from Europe to the US. India, the second-most populated country in the world, has adopted RFID technology to cope with the rapid increase of volume and types of waste (Infotech 2013). Similarly in 2010, what China faced were the World Expo and huge amounts of construction waste that comprised 30 % to

40 % of the total urban waste. Shanghai was chosen for a pilot project using an RFID based waste management system. All the waste dumping trucks had an embed RFID tag and volume of waste they carry was checked by the local government (Ruan and Hu 2011). Another interesting case of environmental application emerges from South Korea. The South Korean government operates U-Street Trees Systems through which the exact location and status of street trees can be monitored. Information about location and status of street trees are collected by an RFID tag that is attached to each tree is saved in a web information system, so trees can be managed effectively. Kim et al. (2006) claim that this web based information system could manage information remotely with an interactive system (Table 4).

Table 4 RFID applications in defense and security

Application	Paper	Year	Case	Country
Waste Management	Kang et al.	2012	Ministry of Environment in the ROK introduced RFID system in the medical waste management in 2006	ROK
	Infotech	2013	India adopted RFID technology in order to deal with rapid increase in volume and types of waste	India
	Ruan and Hu	2011	Shanghai city government started RFID using waste management to prepare Shanghai expo 2011	China
	WWICS	2008	How US government using RIFD in waste	US

Application	Paper	Year	Case	Country
			management	
	Schindler et al.	2012	Cases of using RFID in waste management in EU member country	EU
	Ransford et al.	2012	Waste management operating system broadly used within US	US
Street tree Management	Kim et al.	2006	ROF government using RFID technology to manage street tree condition	ROK

3.4 Transportation

Public transportation is another popular sector for RFID technology applications. RFID based electronic toll collection technology is one of the oldest and widespread RFID implementation (Ulatowski 2007). As soon as an RFID tag embedded car arrives at a toll booth, the RFID reader scans and reads the information that the RFID tag contains. The driver will pay debits according to the price that electronic reader suggests. In the US, electronic toll collection is thought as efficient and effective method that eliminates long lines of traffic at toll booth (Ulatowski 2007). RFID based toll collection is also adopted in criminal cases because it enables prosecutors to identify the exact location of the criminal's car (Smith 2006). In South Korea, the Korean government has set credit card-linked electronic toll collection system called 'Hypass' especially for collecting transportation tolls on express ways. If an RFID tag is embedded on their cars, drivers can pass the tollbooth without stopping the car because RFID reader scan the data immediately and handle the whole payment process in about 5 s (Kim 2008). Hong Kong launched similar public transportation toll collection system in 1997 and the 'Octopus Card' is now internationally famous for its convenience. This system is able to handle 10 million transactions per day and includes all modes of public transport (Kovavisaruch and

Suntharasaj 2007). South Korea has set credit card-linked electronic toll collection system called 'Hypass' especially for collecting transportation tolls on express ways. RFID technology is also implemented in railroad toll collection in India, where railroads are the most widely used form of public transportation. If an RFID tag is embedded on their cars, drivers can pass the tollbooth without stopping the car because RFID reader scan the data immediately and handle the whole payment process in about 5 s (Kim, 2008). In addition, RFID has been used as a critical technology to promote efficiency and transparency for public transportation system in developing countries. For instance, the Mexican government runs "Creating Traffic Knowledge in Mexico: Applying RFID to prevent vandalism" and one of the purposes of this innovative project is to develop a transportation information system to acquire more subtle data necessary for government decision making (Prado et al. 2010). Analogous to Mexican case, in Bangladesh where BRTA (Bangladesh Road Transport Authority) was started in 2003, the technology is operated mainly for control and supervision of the road transport systems (Hossain et al. 2009). RFID technology is also implemented in railroad toll collection in India, where railroads are the most widely used form of public transportation (Table 5).

Table 5 RFID application in transportation

Application	Paper	Year	Case	Country
Public Transportation	Ulatowski	2007	Electronic toll collection	US
	Kovavisaruch and Suntharasaj	2007	Well-known success smart card case in Hong Kong	Hong Kong
	Kovavisaruch and Suntharasaj	2007	Largest smart card transport system in the world	China

Application	Paper	Year	Case	Country
	Pransanth and Soman	2009	RFID based transportation system especially for railroad toll collection in India	India
	Kim	2008	Hypass case used in highway toll collection in ROK	ROK
Managing Road condition	Prado et al.	2013	How Mexican government using data collected from RFID system in decision making	Mexico
	Hossain et al.	2012	Monitor and control the road system	Bangladesh

3.5 Healthcare and welfare

RFID enables hospitals to manage their equipment more easily and save expenses in public health areas[12]. The US government agencies like FDA have also already used RFID tag in monitoring drug industry[13]. Since American hospitals handle almost 4,000 medicines per day, medication errors can be easily occurred. With strong government support, public hospitals in Taiwan have actively adopted innovation of RFID (Kuo and Chen 2008)[14]. Even though it is not yet commercialized, an RFID identification system[15] for the visually impaired people is being developed by engineers in Pakistan with the support of the Pakistani government (Murad et al. 2011) (Table 6).

Table 6 RFID application in healthcare and welfare

Application	Author	Year	Case	Country
Managing Public Hospital	Kuo and Chen	2008	Many hospitals are actively involved RFID system in managing hospital with the support of the government	Taiwan
Pharmacy	Wyld	2008	U.S. government and FDA recommended that pharmaceutical industries move to implement RFID tag to prevent counterfeit dugs by 2007	US
	Thuemmler et al.	2007		
	Romero and Lefebvre	2013		
	Skinar	2005	Florida state government imposed fine to drug suppliers when they did not adopt RFID tags	Florida State
Service for the impaired	Murad et al.	2011	RFID tag using service for visually impaired people was designed and implemented by	Pakistan

Application	Author	Year	Case	Country
			Pakistan government	
Infection Management	Nicholas	2008	Singapore government track visitors, patients and staff, to Figure out who was the SARS virus bearer in 2003	Singapore
	Kuo et al.	2004	Taiwan hospital proceed RFID plans to track SARS patients	Taiwan

3.6 Agriculture and livestock

RFID technology can be an effective tool for securing food safety and managing agriculture and livestock. Another major advantage to this system is that animal disease tracking can be realized through innovative technology like RFID (Hossain and Quaddus 2009). With the government support, researchers have developed the Navigation System for Appropriate Pesticide Use as a basic system for risk management in agriculture (Nanseki et al. 2005). RFID technology in agriculture was first introduced by the European Union (EU) in the late 1990s and shortly thereafter many countries, such as Australia, Japan and South Korea, adopted the innovation. Among those countries, the Australian government was the most passionate in implementing RFID[16]. For instance, all the livestock in Australia have RFID embedded tags on their bodies immediately after they are born; information that enables farmers to identify each entity and its health status is registered in National Livestock Identification System (NLIS). RFID technology in Japan has been also adopted in agriculture especially to secure food safety and agricultural risk management that can occur by abusing pesticides (Nanseki et al. 2005, Sugahara 2009). The Japanese government planned to make a food traceability system by 2010 as a part of the "e-Japan" plan (Chen et al. 2008). The United States is another case that applies mandatory RFID based identification system in managing livestock. According to RFID Gazette (2006), the USDA is pushing for RFID tagging of cattle to make tracing of disease patterns easier. With the formation of National Institute for Animal Agriculture (NIAA) in 2002, the plan for setting the National Animal Identification System was started. What the US government fulfilled through this program was "to be able to identify all animals and premises that have had contact with a foreign or domestic animal disease of concern within 48 h" (Wyld, 2005) because "the sooner animal health officials can identify infected and exposed animals and premises, the sooner they can contain the disease and stop its spread (USDA-APHIS 2005)" (Table 7).

Table 7 RFID application in Agriculture and Livestock

Application	Paper	Year	Case	Country
Agricultural risk management	Nanseki et al.	2005	Navigation system for Appropriate Pesticide Use was developed as a system for agricultural risk management	Japan
	Sugahara	2009		
Livestock Management and Disease Tracking	Trevarthen and Michael	2007	Cochrane Dairy Farm let their cows in herd have National Livestock Identification System	Australia
	Hossain	2009	How Australian farms adopt and implement RFID technology	Australia

Application	Paper	Year	Case	Country
	Wyld	2008	US began National Animal Identification System in 2002	US

3.7 Public policy issues from RFID diffusion

RFID applications and diffusion generate complex policy and governance problems. We address public policy issues such as technological gap and uncertainty of expecting potential benefits and costs from a rapid and massive RFID diffusion. Uneasy governing issues in transparency, digital identification and power distribution are arising from inappropriate RFID applications. We discuss governance issues such as corruption, privacy problem, and digital monopoly and literacy in the following.

3.8 Technological concerns

Technology is not still enough to satisfy all the elements that RFID is trying to perform various operational mechanisms. RFID technology deficiencies inevitably occur with the application of technology because there is niche space still left. For instance, RFID technology does not have a unified frequency standard yet. Since there are no internationally agreed upon frequencies for RFID operations, permitted scanner/reader powers also differ between countries. There are still significant differences between the frequencies from the EU and the USA (Hossain et al., 2009). In addition, Reichenhach (2008) pointed out the lack of storage capacity. In the EU, where RFID based waste management is common, there are technological barriers like a shortage of storage capacity[17]. Ema and Fujigaki (2011) draw implications from a child monitoring case done in Japan that being informed of children's exact location cannot guarantee their actual safety, but RFID tags often lead to that cherished illusion. Vining (2005) warned about another possibility of niche space. According to his study about port security in the US, stealing goods without damaging RFID tag is possible because at ports, the container can be drilled into and contents can be removed. The RFID tag does not have to endure any damage through this whole process. In the US, as a response to continued pressure from various stakeholders, the US government even adopted the 'Faraday cage' for privacy protection[18] (Table 8).

Table 8 Technological Issues

Issue	Paper	Year	Case	Country
Cryptography	Laurie	2007	Cryptography is not a perfect technique for protecting information saved in RFID tag so it can be easily attacked by the hackers	US
Standard	Hossain et al.	2009	No internationally agreed frequencies for RFID operations	
Storage Capacity	Reichenbach	2008	There is shortage of capacity for source-separated waste in household	Europe
NicheSpace	Ema and Fujigaki	2011	Informing children's exact location cannot guarantee their actual safety	Japan
	Vining	2005	In port, the container can be drilled into and contents can be	US

Issue	Paper	Year	Case	Country
			removed without disturbing the RFID tag	
	Ezovski et al.	2007	Without any additional protections, the Faraday cage is not safe enough	US

3.9 Uncertain cost-benefit effectiveness

RFID defenders emphasize that RFID technology can guarantee effectiveness and efficiency at a very cheap price. There is, however, substantial evidence to show RFID can generate unexpected costs[19]. In reality, the RFID tag is much more expensive than a barcode, which was very popular in identifying materials before the rise of RFID technology (Becker 2004). Purchasing RFID devices, hardware, and tags is not sufficient to drive system relevantly. To guarantee a better quality of service, the RFID system needs more additional things such as "circular process mechanism, the richness of consultant, project manager, programmers and plentiful project labors" (Kuo and Chen 2008). These elements for a better RFID performance may involve considerable costs. Kuo and Chen (2008) reported that RFID technology consumers and government should pay the extra hidden cost in the healthcare industry (Table 9).

3.10 Dubious transparency and corruption

RFID technology is expected to increase transparency and monitor corruption. However, RFID technology cannot ensure a high level of transparency than expected. As a matter of fact, RFID tags can be cloned and manipulated quite easily, and this kind of tag corruption can occur at every stage of RFID implementation. There are various examples to show an inappropriate use of RFID technology. For instance, Armknecht et al. (2010a, b) warned the possibility of tag corruption. Lee et al.

(2012) pointed out reader corruption of the RFID technology. An existing security model mainly focuses on the possibility of tag corruption, but reader corruption can hurt consumers' privacy as seriously as tag corruption can. Jules (2006) reported one of the tag corruption cases that observed in United States. One of the staff members who worked in a Dupyu store, an unscrupulous retailer, attached a cloned tag to counterfeit drugs. Avoine et al. (2010) argued that internet based databases can also be directly attacked and emphasized the possibility of reader corruption. There are also unethical behaviors to avoid RFID monitoring process. In the EU where an RFID based waste management system is aggressively implemented, some people disposed waste that came from their house at work places in order to avoid exact calculation through the RFID system. Not only this, Bilitewski (2008) reported that some conscienceless people are burning or transferring waste outside instead of throwing it into their RFID tag attached garbage can (Table 10).

Table 9 Cost-benefit effectiveness issues

Issue	Paper	Year	Case
Cost	Becker	2004	RFID tag is much more expensive than barcode
Hidden Cost	Kuo and Chen	2008	There is still hidden cost for highly qualified RFID system
Privacy	Jensen	2008	Active tag, which is more safe for securing privacy than passive tag is more expensive than passive tag

3.11 Privacy issues

One of the most serious issues that RFID technology faces today is whether RFID technology is secure enough to protect privacy. Privacy is the most important concern RFID users have to deal with (Perakslis and Wolk 2005). RFID tag embedded chips often contain important personal information and usually this kind of private information can hurt one's privacy seriously if leaked. To prevent leakage of private information, engineers developed cryptography, but there remains criticism[20]. The reason why these sorts of privacy concerns arise is because of the lack of security protection capacity of modern RFID technology. As we discussed above in the technological issues section, RFID technology today is not developed to secure perfect privacy. The technology itself has lots of deficiencies and people are smart enough to find niche spaces that can destroy the RFID security process. RFID itself can involve not only various hidden costs[21] but also induces a serious privacy problem[22]. However, despite these possibilities of attacks on privacy, there are lots of stakeholders and scholars who advocate the potential benefits of RFID. They claim that tracking and profiling consumers is solely for implementing RFID chips more effectively. Eaward Rerisi, one of the producers of early implementation of RFID technology argued that, "An RFID reader can read the number on a tag, but without knowing what the number means, there is no way to access personal information. The idea that the tags can be read by just anybody—that's pretty impossible" (Murray 2003) (Table 11).

Table 10 Transparency issues

Issue	Paper	Year	Case	Country
Tag Corruption	Armknecht et al.	2010	Tag corruption can be occurred at every stage of RFID implementation	
	Jules	2006	Dupyu staff attach cloned tags to counterfeit drugs	US
Reader Corruption	Lee et al.	2012	Although reader corruption can cause serious privacy attack, most of the scholars do not consider it as security problem.	

Issue	Paper	Year	Case	Country
Cheating	Reichenbach	2008	Waste being disposed of at work places	EU
	Bilitewski	2008	Burning waste or transferring it to outside	EU

Table 11 Privacy issues

Paper	Year	Category	Contents
Hwang et al.	2009	Cloning	The attacker can read the tag and then clone the tag by writing all the obtained data into other
		Eavesdropping	The attacker surreptitiously listens to all the communications between the reader and the tag
		Replay attack	The attacker repeats or delays the same message when valid data are transmitted
		Denial of Service	The attacker can send massive message to RFID system and attempt to crash the RFID system
		Forward Security	The attacker can compromise a tag and obtain its current relation date

Paper	Year	Category	Contents
		Tag tracing	The tag always broadcasts a fixed serial number to somewhere nearby the reader; therefore, the adversary can identify a fixed serial number of the tag from different locations or transaction records
		Individual data privacy	The hacker can know what items the consumer bought from the store or what books the consumer borrowed from the library
		Data forging	The attacker can modify the dates, items, and prices and then cause great loss if the tag can store extra data
Numann and Hogben	2008	Skimming	The attacker opens a clan-destine connection to the chip and gains access to the data
		Eavesdropping	The attacker intercepts the communication between the chip and an authorized reader
		Location tracking	The attacker generates person or card-specific movement profiles.

3.12 Unequal power and digital literacy

Unequal distribution of RFID technology can generate unequal distribution of various resources such as information and digital literacy. Especially in developing countries, the combination of unbalanced power distribution between stakeholders and a low level of digital literacy can cause serious problems. Ketprom et al. (2007) emphasized that in developing countries like Thailand[23], governments should provide education and training on how to use brand new technology to poor farmers whose digital literacy remains relatively low. But poor farmers in Thailand are not the only stakeholders who are suffering from a lack of digital literacy. In Bangladesh, where RFID toll collection is common, traffic policies have no interest in using RFID technology for managing public transportation systems. Rather, they prefer traditional ways of toll collecting to information based technology (Hossain et al., 2009). Prasanth et al. (2009) found that the lack of digital literacy among the Indian people hampered an effective process of railroad toll collection in India. Another problem developing countries face is an unbalanced power distribution due to lack of democratic value embedded governance. Chen et al. (2008) criticized the Taiwanese government because it monopolizes most of the information collected by RFID technology. As we already discussed above, when RFID tag scanned, information saved in RFID tag is scanned by reader and then transmitted to an internet based database. If that data were available to the public, individuals and industry could make more reasonable decisions by analyzing them. We find another unbalanced power distribution case in China's waste management system. According to Ruan and Hu (2011), the Chinese government benefits most from the RFID system[24] (Table 12).

Table 12 Power distribution and digital literacy issues

Issue	Paper	Year	Case	Country
Digital Literacy	Ketprom et al.	2007	RFID technology is widening the gap between poor and rich farmers	Thailand
	Hossain et al.	2009	Traffic policies are not interested in using RFID technology	Bangladesh

Issue	Paper	Year	Case	Country
Power Distribution	Ruan and Hu	2011	Price of construction waste transportation is set by government, so other stakeholders have no choice	China
	Chen et al.	2008	Consumer can get only limited information	Taiwan

4. DISCUSSION AND CONCLUSION

We found, relying on a systematic review from 111 RFID studies, six key areas of RFID applications. Specifically in the defense and security section, we addressed how military and airports/ports manage RFID systems to ensure security. We also found that RFID is effectively implemented in prison management and child protection programs. Numerous governments have introduced RFID identification tools such as e-passport and e-ID. RFID systems for waste management and street tree management are widely used from rich to poor countries. In healthcare and welfare delivery, RFID based smart cards have turned out to be very efficient. RFID is now being used to monitor counterfeit drugs. RFID has been applied to delivering service for the impaired and to trace infection. However, despite potential benefits from RFID applications, various unexpected problems arise. RFID can still involve technological deficiencies, especially in securing cryptography techniques, international standards of frequency, and storage capacity. RFID technology is not still enough to be efficient and effective in some areas (Becker 2004, Jensen et al.2007). Tag and reader corruption can hurt transparency and security. Privacy issues are still the most serious issues that RFID faces today (Naumann and Hogben 2008). RFID itself can generate new unequal digital literacy and power distribution, especially in developing countries such as Thailand and Bangladesh. Even the most latest innovative technologies, like RFID, do not have perfect answers to securing efficiency, effectiveness, convenience, and transparency. Rather, RFID technology itself creates unexpected problems. It should be noted that democratic governance and trust is still important to technological innovation and policy issues arising from a rapid RFID diffusion.

Our systematic review is incomplete to discuss all of the RFID issues from technology, market and management, e-government, and legal aspects. Further research on RFID diffusion and impact include not only various theoretical issues of but also legal and managerial problems. For instance, both qualitative and quantitative research is required to explore what factors are critical to adopt and implement new RFID technology in terms of governance and digital literacy. Both micro and macro approaches with massive data are also required to identify how RFID improve not only organizational performance in government agencies and various industry sectors but also quality of our life.

FOOTNOTES

1 For example, after serious attack by Osama Bin Laden on 9/11, the American government decided to implement an RFID tag embedded e-passport and VISA waiver program. The US government asked their member countries to implement e-passport by late 2005 and soon US member countries like ROK and EU started to use e-passports. Currently, no one can enter to United States without an RFID tag based e-passport.

2 Especially in developing countries, governments usually adopt brand new IT technologies, but their low level of socio-economic infrastructures may prohibit the efficient operation of technology.

3 For instance, RFID applications may lack social virtues like trust, ethics, and democracy. It is essential to understanding how a rapid diffusion and massive applications of RFID generate conflict or harmony among human behaviors, digital literacy, institutional rules, and technology.

4 US Army and its allies could not only manage weapons and soldiers but also identify who was the enemy or not (Castro and Wamba, 2007). This whole project of developing an RFID based identifying system was known as IFF (Identify Friend of Foe).

5 See more various examples of RFID applications at the website (https://epic.org/privacy/rfid/)

6 In 2004, the US Army adopted RFID during the Iraq war to track Iraq troops. Not only these, the US Army piloted 4 projects using RFID; identifying material locations, weapons deteriorations, hazardous material tracking, and asset tracking (Anon 2002).

7 The New York City government also started an RFID e-seal pilot project in the New Jersey Port. Once RFID read and scan the tag, it can identify the contents of the container box. Also, the port of Tacoma and Seattle planned to adopt E-Seals, made of metal bolts with embedded RFID devices to ensure its security (Konsynski and Smith 2003).

8 Calpatria prison, located in Los Angeles, adopted a prisoner monitoring system using RFID chips in 2000 as the very first pilot using RFID in prison management in United States (Kim 2008). According to regulations of Calpatria prison, all the prison inmates were issued bracelets in which RFID chips were embedded. Since the pilot project at the Calpatria prison was successful, the local government let other prisons in Los Angeles adopt the innovative bracelet. The LA County prison started to use a brand new bracelet in response to the state government's order; it is reported that through its adoption the prison could increase efficiency and effectiveness and decrease crimes occurred between inmates simultaneously (Nicholas 2008).

9 For instance, city governments in the Gifu and Osaka prefecture provided RFID tags that can be attached in students' schoolbags to public elementary schools (Ema and Fujigaki, 2011). Similarly, in Haewoondae beach, one of the most famous vacation spots in South Korea, Busan Metropolitan City provides for parents RFID embedded bracelet that enables tracking exact location of their child by a smart phone (http://news.naver.com/main/read.nhn?mode=LSD&mid=sec&sid1=102&oid=001&aid=0003353674).

10 This project began in 2002, but it took 3 years to fully implement for all 16 US passports (Meingast et al. 2007). The appearance of the e-passport is very similar to old passports, but woven into the paper of the passport, there is RFID tag that information about owner of the passport is included. Information about nationality, sex, age, and so on is scanned, as airport staff members scan the passport through RFID reader (Lorenc 2007). The US government did not stop at this point and adopted VISA Waiver Program. By 2005, the US member countries had to adopt RFID based e-passports and VISA Waiver Programs in order for their citizens to enter the United States because without e-passports, passengers could not be accepted at American points of entry. Today, the e-passport includes not only individual data but biological data, such as fingerprints (ICAO TAG MRTD/TWF 2004).

11 E-Government News, "EU Asks US for Time to Issue Biometric Passports". iDABC European e-Government Services, 1 April, 2005.

12 For instance, Wicks et al. (2006) reported that this RFID based hospital management system is very effective in reducing management costs because embedded RFID tags can track lost or hidden expensive equipment. Miller (1999) pointed out that the potential of RFID technology can be expanded in tracking the location of patients and controlling the drugs. Also, Chowdhury and Khosla (2007) argued that RFID technology can be effectively used not only in hospital equipment management but also in patient management.

13 According to statistics published by the Institute of Medicine (IOM), about 44,000 to 98,000 people die in the USA per year because of improper drug administration (Kohn et al. 2000). To rectify this phenomenon, in 2004 the US government and US FDA recommended pharmaceutical industries to implement RFID tags to prevent the production of counterfeit drugs (Wyld, 2005). The Florida state government added legal regulations to this recommendation in 2006. If a pharmaceutical industry located in Florida does not attach RFID embedded tags on its products, it has to face substantial financial penalties (Skinar 2005).

14 For instance, RFID technology saved Singapore and Taiwan from SARS attack around 2003. Two public hospitals in Singapore in 2003 adopted RFID technology to track staff, patients, and visitors in order to trace people who carried the SARS virus. The RFID information database saved all the data collected from each individual's RFID tag for 21 days, which was thought to be long enough for expression of SARS virus (Nicholas 2008). A similar process was done in Taiwan too. During the SARS period, five hospitals, including Taipei Veteran's General Hospital, implemented RFID tags to track patients who had possibility contracted the SARS virus (Kuo et al. 2004) with strong government support.

15 One peculiar characteristic of this system as compared to other systems is that visually impaired people are both the reader carrier and the service beneficiaries, simultaneously. Generally in government service delivery using RFID technology, the service provider usually carries RFID tag readers and service beneficiaries usually act more passive roles by attaching RFID tags. But in this Pakistani case, the service beneficiaries can identify objects around them by operating RFID tag reader they have.

16 The Australian government passed legislation on mandatory use of RFID tags in the livestock industry, so Hossain and Quaddus (2014) categorized Australian case as a very rare and special adoption case. According to Trevarthen and Michael (2007)'s case study, one of the Australian farms where the RFID tag is implemented, farmers not only track the exact location of the cows but also check the condition, identify cows and even feed the new born cows automatically by using an RFID system.

17 Usually people throw waste in various places. They may throw it away at their houses or at work places, like an office. Since the RFID tag is only attached to a garbage can in house, it is impossible to track all types of waste throwing behaviors. Inevitably, this shortage of storage capacity leads to selective waste collection monitoring.

18 The faraday cage is an object in metal; proponents of this device argue that faraday cage can prevent hacker's attack because electronic devices are prevented from passing through the object (Ezovski et al. 2007). But speculation about stability of this technology still remains. Lorenc (2007) reported that if there is no additional technology, the faraday cage cannot preserve sensitive security.

19 As a matter of fact, there are two kinds of RFID tags, the passive tag and the active tag. These two tags provide owners with different benefits and liabilities. Active tags are implemented by a power source, such as small battery. Active tags are more efficient and safe in protecting privacy than passive tags, so sensitive organizations like military prefer active tags to passive tags. But ordinary consumers have less accessibility to active tags because active tags are much more expensive than passive ones (Jensen et al. 2007).

20 According to Laurie (2007), even without physically losing an RFID tag, private information can be stolen because what's inside RFID tags can be skimmed quite easily. If we can build a device that enables us to transmit an arbitrary number, invading the internet based database is theoretically possible.

21 For instance, Wal-Mart Stores Inc. and Procter & Gamble Co. in autumn 2003 planned a very interesting experiment to check the potential deficiencies of RFID technology. Customers of a Wal-Mart store in Broken Arrow, Oklahoma were secretly tracked through RFID tags while purchasing lipstick that Procter & Gamble Co. made (Barut et al. 2006).

22 For instance, Hwang et al. (2009) and Numann & Hogben (2009) categorized various kinds of privacy attack cases that can possibly occur. According Hwang et al. (2009), technological deficiency enables hackers to engage in cloning, eavesdropping, replay attack, denial of service, forward security, tag tracing, individual data privacy, and data forging. Specifically, the hacker can read the tag and then clone the tag (cloning), surreptitiously listen to all the communications between and the tag (eavesdropping), repeat or delay the message (replay attack), send large amount of message to break down RFID system (denial of service), compromise a tag (forward security), trace the exact location of the tag (tag tracing), find out shopping trend of the consumer (individual data privacy), and modify information saved on an RFID tag (Data forging). In addition, Numann and Hogben (2008) categorized the privacy attacking cases more briefly. According to their research, the hacker can attack RFID tag in some ways. First, the attacker can open a connection to the chip and can steal the data inside (skimming). Second, the attacker can intercept the communication between tag and reader (eavesdropping). Last, the attacker can track the exact location of the tag or the person.

23 In Thailand, most of the farms are trying to adopt an RFID system in farm management, but RFID technology is widening the gap between poor and rich farmers. Poor farmers are usually less educated people who have hardly had any experience using digital technology like RFID. On the other hand, well-educated wealthy farmers face low entry barriers and easily adopt the technology. Rich farmers armed with innovative technology not only make enormous fortunes by increasing efficiency but also by replacing poor labors with RFID embedded devices. According to the Thailand Ministry of Labor (2006), most farm labors are afraid of being replaced. Unfortunately, this phenomenon eventually correlates to a serious gap between rich and poor.

24 The government has invested a huge amount of money to buy necessary devices such as readers, tags, hardware, and so on, but in the long run cost of management will decrease. On the other hand, the waste industry could carry a very heavy debt. The waste contractors have to deal with expensive RFID tag rental and as well as the cost of construction simultaneously. Although this situation is totally unfavorable for them, the waste management industry cannot resist to this policy because the Chinese government is the entity which has made the use of RFID policy and set prices. The industry has no other choice but consent.

ACKNOWLEDGMENTS

This paper was supported by the research grant of Seoul National University Foundation (Korea Institute of Public Affairs) in 2015.

COMPETING INTERESTS

The authors declare that they have no competing interests.

AUTHORS' CONTRIBUTIONS

KJ and SL carried out a systematic literature review not only from not only technological point of view but also from social scientific point of view. Both authors read and approved the final manuscript.

REFERENCES

1. Armknecht F, Chen L, Sadeghi AR, Wachsmann C. Anonymous authentication for RFID systems. In: Radio Frequency Identification: Security and Privacy Issues. Berlin Heidelberg: Springer; 2010a. p. 158–75.

2. Armknecht F, Sadeghi AR, Visconti I, Wachsmann C. On RFID privacy with mutual authentication and tag corruption. In: Applied Cryptography and Network Security. Berlin Heidelberg: Springer; 2010b. p. 493–510.

3. Avoine G, Coisel I, Martin T. Time measurement threatens privacy-friendly RFID authentication protocols. In: Radio Frequency

Identification: Security and Privacy Issues. Berlin Heidelberg: Springer; 2010. p. 138–57.

4. Barut M, Brown R, Freund N, May J, Reinhart E. RFID and corporate responsibility: hidden costs in RFID implementation. Bus Soc Rev. 2006;111:287–303.

5. Becker C. A new game of leapfrog? RFID is rapidly changing the product-tracking process. Some say the technology--once costs drop--could displace bar-coding. Modern Healthcare. 2004;34:38–40.

6. Bilitewski B. From traditional to modern fee systems. Waste Management. 2008;28(12):2760–6.

7. Castro L, Wamba SF. An inside look at RFID technology. J Technol Manag Innov. 2007;2:128–41.

8. Chen RS, Chen CC, Yeh KC, Chen YC, Kuo CW. Using RFID technology in food produce traceability. WSEAS Transac Inform Sci Appli. 2008;5:1551–60.

9. Chowdhury B, Khosla R. RFID-based hospital real-time patient management system. In: Computer and Information Science, 2007. ICIS 2007. 6th IEEE/ACIS International Conference on. 2007. p. 363–8. IEEE.

10. Ema A, Fujigaki Y. How far can child surveillance go?: Assessing the parental perceptions of an RFID child monitoring system in Japan. Surveillance Soc. 2011;9:132–48.

11. Ezovski GM, Watkins SE. The electronic passport and the future of government-issued RFID-based identification. In: RFID, 2007. IEEE International Conference on. Grapevine, TX: IEEE; 2007. p. 15–22.

12. Hossain MA. Exploring the Perceived Measures of Privacy: RFID in Public Applications. Aus J Inform Systems. 2014;18(2):133–48.

13. Hossain MA, Quaddus M. An adoption-diffusion model for RFID applications in Bangladesh. In: Computers and Information Technology, 2009. ICCIT'09. 12th International Conference on. 2009. p. 127–32. IEEE.

14. Hossain MF, Sohel MK, Arefin AS. Designing and implementing RFID technology for vehicle tracking in Bangladesh. Dhaka, Bangladesh:

National Conference on Communication and Information Security; 2009.

15. Hossain MA, Quaddus M. Developing and validating a hierarchical model of external responsiveness: A study on RFID technology. Inform Systems Frontiers. 2014;17(1):109–25.

16. Hwang MS, Wei CH, Lee CY. Privacy and security requirements for RFID applications. J Comput. 2009;20:55–60.

17. Jensen A, Cazier J, Dave D. The impact of government trust perception on privacy risk perceptions and consumer acceptance of residual RFID technologies. AMCIS 2007 Proceedings, 146. (2007).

18. Jensen AS, Cazier JA, Dave DS. Mitigating consumer perceptions of privacy and security risks with the use of residual RFID technologies through governmental trust. J Inform Syst Security. 2008;4:41–66.

19. Jules A. RFID security and privacy: A research survey. Selected Areas Commun IEEE J. 2006;24:381–94.

20. Infotech. RFID based Waste Management System. 2013. http://www.slideshare.net/iaitoinfotech/rfid-in-waste-management-slide-share.

21. Ketprom U, Mitrpant C, Lowjun P. Closing digital gap on RFID usage for better farm management. In: Management of Engineering and Technology, Portland International Center for. Portland, OR: IEEE; 2007. p. 1748–55.

22. Kim JG. A divide-and-conquer technique for throughput enhancement of RFID anti-collision protocol. Communications Letters, IEEE. 2008;12:474–6.

23. Kim EM, Pyeon MW, Kang MS, Park JS. A management system of street trees by using RFID. In: Web and Wireless Geographical Information Systems. Berlin Heidelberg: Springer; 2006. p. 66–75.

24. Konsynski B, Smith HA. Developments in practice x: Radio frequency identification (rfid)-an internet for physical objects. Commun Assoc Inform Systems. 2003;12:19.

25. Kovavisaruch L, Suntharasaj P. Converging technology in society: opportunity for radio frequency identification (RFID) in Thailand's

transportation system. In: Management of Engineering and Technology, Portland International Center for. Portland, OR: IEEE; 2007. p. 300–4.

26. Kuo CH, Chen HG. The critical issues about deploying RFID in healthcare industry by service perspective. In: Hawaii International Conference on System Sciences, Proceedings of the 41st Annual. Waikoloa, HI: IEEE; 2008. p. 111–1.

27. Kuo F, Lee Y, Tang CY. The Development of RFID in Healthcare in Taiwan. Bejing: ICEB; 2004. p. 340–5.

28. Laurie A. Practical attacks against RFID. Network Security. 2007;2007:4–7.

29. Lee K, Nieto JG, Boyd C. A state-aware RFID privacy model with reader corruption. In: Cyberspace Safety and Security. Berlin Heidelberg: Springer; 2012. p. 324–38.

30. Li S, Visich JK, Khumawala BM, Zhang C. Radio frequency identification technology: applications, technical challenges and strategies. Sensor Review. 2006;26:193–202.

31. Lorenc ML. Mark of the Beast: US Government Use of RFID in Government-Issued Documents. The Alb LJ Sci Tech. 2007;17:583.

32. Meingast M, King J, Mulligan DK. Embedded RFID and everyday things: A case study of the security and privacy risks of the US e-passport. In: RFID, 2007. IEEE International Conference on. Grapevine, TX: IEEE; 2007. p. 7–14.

33. Murad M, Rehman A, Shah AA, Ullah S, Fahad M, Yahya KM. RFAIDE—An RFID based navigation and object recognition assistant for visually impaired people. In: Emerging Technologies (ICET), 2011 7th International Conference on. Islamabad: IEEE; 2011. p. 1–4.

34. Murray C. Privacy concerns mount over retail use of RFID technology. EE Times. 2003;2:4–5.

35. Nanseki. A navigation system for appropriate pesticide use: design and implementation. Agricultural Information Research. 2005;14:207–26.

36. Naumann I, Hogben G. Privacy features of European eID card specifications. Network Security. 2008;8:9–13.

37. Perakslis C, Wolk R. Social acceptance of RFID as a biometric security method. In Technology and Society, 2005. Weapons and Wires: Prevention and Safety in a Time of Fear. ISTAS 2005. Proceedings. 2005 International Symposium on (pp. 79-87). IEEE. New York, New York; 2005.

38. Prado JAD, Monterrey IC, Prado FED. Creating Traffic Knowledge System in Mexico: Applying RFID to Prevent the Vandalism. The 15th International Business Information Management Association Conference. 2010. p. 2042–50.

39. Prasanth V, Hari PR, Soman KP. Ticketing Solutions for Indian Railways Using RFID Technology. In: Advances in Computing, Control, & Telecommunication Technologies, 2009. ACT'09. Trivandrum, Kerala: International Conference on. 2009. p. 217–9. IEEE.

40. Ransford B, Sorber J, Fu K. Mementos: system support for long-running computation on RFID-scale devices. Acm Sigplan Notices. 2012;47:159–70.

41. Reichenbach J. Status and prospects of pay-as-you-throw in Europe–A review of pilot research and implementation studies. Waste Management. 2008;28:2809–14.

42. Romero A, Lefebvre E. Gaining Deeper Insights into RFID Adoption in Hospital Pharmacies. World. 2013;3:164–75.

43. Ruan T, Hu H. Application of an RFID-based system for construction waste transport: a case in Shanghai. In: Computational Logistics. Berlin Heidelberg: Springer; 2011. p. 114–26.

44. RFID Gazette. RFID hybrid tech: Combining GPS for location tracking. Retrieved December 5, 2006, from tec.html http://www.rfidgazette.org/2006/12/index.html.

45. Schindler R, Schmalbein N, Steltenkamp V, Cave J, Wens B, Anhalt A, et al. SMART TRASH – study on RFID tags and the recycling industry, technical report. TR-1283-EC. Santa Monica, CA: RAND Europe Corporation; 2012.

46. Shahram M, Manish B. RFID Field Guide: Deploying radio frequency identification systems. New York: Prentice Hall; 2005.

47. Smith JE. You Can Run, But You Can't Hide: Protecting Privacy from Radio Frequency Identification Technology. NCJL Tech. 2006;8:249.

48. Sugahara K. Traceability system for agricultural productsbased on RFID and mobile technology. In: Computer and Computing Technologies in Agriculture II, vol. 3. US: Springer; 2009. p. 2293–301.

49. Tien L. RFID tags should track inventory, not people. RCR Wireless News, 2004. http://www.rcrwireless.com/20040705/archived-articles/rfid-tags-should-track-inventory-not-people.

50. Takahashi. "The Father of RFID," Mercury News, www.siliconvalley.com (June 7, 2004).

51. Trevarthen A, Michael K. Beyond mere compliance of RFID regulations by the farming community: a case study of the Cochrane dairy farm. In: Management of Mobile Business, 2007. ICMB 2007. International Conference on the. Toronto, Canada: IEEE; 2007. p. 8–8.

52. Tsai FMC, Huang CM. Cost-Benefit Analysis of Implementing RFID System in Port of Kaohsiung. Procedia-Social Behav Sci. 2012;57:40–6.

53. Ulatowski LM. Recent developments in RFID technology: weighing utility against potential privacy concerns. ISJLP. 2007;3:623.

54. USDA APHIS. National Animal Identification System: Animal Identification Number(AIN). Retrieved from the web on June 8, 2005. Available at https://www.aphis.usda.gov/traceability/downloads/NAIS_overview_report.pdf.

55. Vining J. RFID Alone Can't Resolve Cargo Container Security Issues. 2005.

56. Weinstein R. RFID: a technical overview and its application to the enterprise. IT professional. 2005;7:27–33.

57. Werb J, Sereiko P. More than just tracking. Frontline Solutions. 2002;3:11–42.

58. Wicks AM, Visich JK, Li S. Radio frequency identification applications in hospital environments. Hosp Top. 2006;84:3–9.

59. Wyld DC. Delta airlines Tags Baggage with RFID. Re:ID Magazine. 2005;1:63–34.

60. Wyld DC. RFID: The right frequency for government. Washington DC: IBM Center for the Business of Government. IBM Center; 2005.

61. Wyld DC. Death sticks and taxes: RFID tagging of cigarettes. Int J Retail Distribution Manag. 2008;36:571–82.

62. Yonhap news. 2013. http://news.naver.com/main/read.nhn?.mode=LSD&mid=sec&sid1=102&oid=001&aid=0003353674.

63. Zhang R. A transportation security system applying RFID and GPS. J Ind Engr Manag. 2013;6(1):163–74.

3

CHOOSING THE RIGHT RFID-BASED ARCHITECTURAL PATTERN

Michel Simatic[1]

[1] INF Department, Télécom Sud Paris, Évry, France

1. INTRODUCTION

RFID provides a way to connect the real world to the virtual world. An RFID tag can link a physical entity like a location, an object, a plant, an animal, or a human being to its avatar which belongs to a global information system. For instance, let's consider the case of an RFID tag attached to a tree. The tree is the physical entity. Its avatar can contain the type of the tree, the size of its trunk, and the list of actions a gardener took on it.

When designing an RFID-based application, a system architect must choose between three locations to store the information: a centralized database, a database locally attached to the device hold by each user of the application, or the tag itself. Each location leads to an RFID-based architectural pattern[1] -. But how to choose the right architectural pattern? What are the application attributes which must be taken into account in order to make the right choice?

The state of the art does not bring satisfactory answers. Indeed, when an article describes a RFID-based architectural pattern, it does not mention the application attributes which lead to choose this architectural pattern. On the other hand, some books or articles present the qualities of architectural patterns. But they do not take into account specificities of RFID. For instance, EPCglobal provides a standardized answer [2]: the centralized architectural pattern. A mobile device, NFC-enabled for example, reads an identifier on the RFID tag, then sends a message to a server which associates the identifier to the avatar stored in a central database. Thanks to its simplicity, this architectural pattern is used by several applications. But, it requires a global computer network: Such requirement increases operational costs. Moreover, it does not withstand an important number of simultaneous RFID read operations. Thus this pattern does not fit all RFID-based applications. [3] presents the stakes of introducing RFIDs inside an enterprise. But it does not contain any system architecture thoughts. In a survey about RFID in pervasive computing, [4] presents several application examples. Depending on the application, avatars are stored either in a central database or in the tags themselves. But the authors do not give any clues on why an application has chosen to store its avatars in a given location. On the other hand, [1] lists the attributes which must be accommodated in a system architecture. Above all, there are the functionalities which are required from the system. Then, orthogonal to these functionality attributes, there are quality attributes. The authors distinguish system quality attributes (availability, modifiability, performance, scalability, security, testability, and usability), business qualities (time to market, cost and benefits, and projected lifetime), and qualities about the architecture itself (e.g. conceptual integrity). But the authors do not focus on RFID specific features.

So we have analyzed several existing industrial or experimental RFID-based applications. Moreover, we have developed RFID-based applications. From this experience, we identify the relevant attributes to compare RFID-based architectural patterns. We present them in section 2. With these identified attributes and their different aspects, we analyze four RFID-based architectural patterns, used by applications to access the avatar of a tagged entity. In the *centralized architectural pattern*, the mobile device reads an identifier on the RFID tag; then it contacts a server which associates this identifier to the avatar stored in a central database or in a database distributed between several companies [2]. With the *semi-distributed architectural pattern*, each mobile device holds a local copy of a central database associating RFID identifiers to avatars [5]. In the *distributed architectural pattern*, each RFID tag holds the avatar [6]. With the *RFID-based Distributed Shared Memory*, RFID tags hold the avatar and a replica of the avatar of other tags [7]. Sections 3 to 6 detail all of these architectural patterns: they present application examples and analyze the architectural pattern with the attributes identified in section 2. Thanks to this analysis, in section 7, we are able to provide guidelines to choose the convenient RFID-based architectural pattern. Finally, section 8 concludes this chapter and proposes perspectives for this work.

2. ARCHITECTURE ATTRIBUTES AND RFID TECHNOLOGY

Relying on the experience gained by analyzing existing RFID-based applications and by developing RFID-based applications, we outline three architecture attributes among the attributes presented in [1]: (i) functionality, (ii) scalability, and (iii) cost. For each attribute, we present its different aspects which are influenced by the use of RFID technology.

2.1. Functionality Attribute

Functionality is the ability of the system to do the work for which it was intended.

All architectural patterns give the ability to read/write the avatar of a read tag.

A first aspect of the functionality attribute is to check how the application behaves when it queries the avatar of a read tag. Is it guaranteed that the returned avatar has indeed the value which was last written? In other words, is there a staleness issue of avatar of a read tag?

The second aspect concerns the possibility of knowing the value (or having an order of idea of the value) of the avatar of a remote tag. By "remote", we mean that the user is not physically near the tag: The user is not able to put her reader on the tag. All she has is the identifier of the remote tag.

The third aspect is the staleness issue of the avatar of a remote tag. If the user is able to know the avatar of a remote tag, is it guaranteed that the returned avatar has indeed the last values associated to the tag?

2.2. Scalability Attribute

The scalability criteria category evaluates how each architectural pattern behaves when there are numerous tags or numerous readers.

Its first aspect is the maximum number of tags which can be handled by the architectural pattern.

The second aspect characterizes the sensitivity of the architectural pattern to the number of simultaneous RFID tag read operations.

2.3. Cost Attribute

The cost attribute groups all of the aspects which have an influence on the installation costs or the operational costs of the RFID-based application.

The first cost aspect concerns the requirement for a global network: do RFID readers have to be able to access at any time and any place to a specific computing machine (for instance, a server in the case of the centralized architectural pattern)? To fulfill this requirement, the readers may be equipped with a wired connection. In that case, the mobility of the readers is limited. The readers may also rely on Bluetooth® or Wi-Fi gateways. Both of these gateways may introduce installation costs. Moreover some readers may not be Wi-Fi enabled. For instance, the Nokia 6212 mobile phone is NFC-enabled, but has no Wi-Fi capabilities. Finally the reader may rely on a mobile data connection (e.g. UMTS, HSDPA, etc.). Such solution introduces operational costs because of data plans.

The second cost aspect concerns the RAM requirement on each tag. The more RAM there is on the tag, the more expensive the tag is. Notice that RAM may actually be prohibited on tags for technical reasons and not for cost reasons. For instance, application may require the use of low-frequency tags (e.g. 125 kHz), so that readers can interact with tags even though there is a liquid between tags and readers. In this case, the throughput is too low for a tag to host information other than its identifier.

The third cost aspect concerns the introduction of a new tag in the environment. For each architectural pattern, we determine the sequence of operations which is required in order to introduce a new tag in the environment. Knowing this sequence, we can determine how long this sequence lasts. Because this initialization procedure is executed by a human or a robot operator, its cost is proportional to the time spent.

The final cost aspect is related to the reinitialization of all of the tags. This criterion concerns only applications which, during their lifetime, need sometimes to have each tag given a new initial value. For instance, this is the case of Paris public transportation system. Users are equipped with a transportation pass containing an RFID tag. At the beginning of a month, each user has to reload her pass (to refresh her access rights): in other words, the tag has to be reinitialized. Some RFID-based games also require tag reinitialization. Indeed, in the case of non-permanent games, users play during successive game sessions. Thus at the beginning of each session, all of the tags must be reinitialized.

In this section, we have defined different aspects of three architecture attributes: (i) functionality, (ii) scalability, and (iii) cost. These aspects are influenced by the use of RFID technology. We use them to compare the behavior of four RFID-based architectural patterns. We start by analyzing centralized architectural pattern.

3. CENTRALIZED ARCHITECTURAL PATTERN

This architectural pattern is often used by manufacturing applications. It has been standardized by EPCglobal [2]. When a reader is near a tag (for instance, the blue mobile in Figure 1), it reads the tag's identifier or an identifier stored in the tag's data zone (its Electronic Product Code in the case of EPCglobal). This identifier is represented by the hexagon in Figure 1. Then, the reader asks a server (ONS lookup service in the case of EPCglobal) which machine (EPC Manager in the case of EPCglobal) manages the avatar corresponding to the read identifier. When the server responds, the reader contacts this machine with the identifier of the tag. The machine queries its database and returns the avatar (for instance, the contents of the hexagon in the database in Figure 1).

Next section gives examples of this architectural pattern.

3.1. Examples

Aspire RFID is an Open Source middleware which is compliant to the specifications of EPCglobal [8]. It proposes several examples of industrial applications for tracking goods.

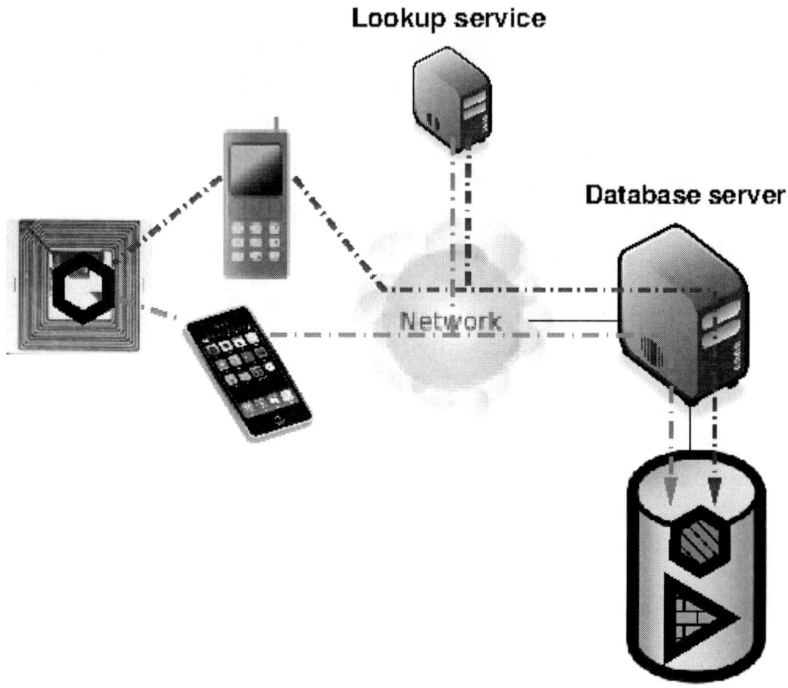

Figure 1. Centralized architectural pattern

Next paragraphs present products or prototypes developed according to centralized architectural pattern, but without being compliant to EPCglobal.

PAC-LAN is a game prototype in which players are equipped with NFC mobile phones without any GPS capabilities [9]. Players must interact with NFC tags which have been disseminated throughout a neighborhood. In a central database, the identifier of each tag is associated to geographical coordinates. When a player reads a tag, her mobile phone uses the UMTS network to contact the server with the tag's identifier. The server queries its database, finds the geographical coordinates, and broadcasts them to all of the players. An administrative application is provided to reset a game on the server. Such reset has an impact on all of the players' mobile phones.

[10] proposes an application so that visitors of an art exhibition can discover the paintings in another way. NFC tags are dispatched on the back of exposed paintings. Equipped with an NFC-enabled phone, the visitor puts her phone on spots of the paintings which intrigue her.

Phone reads the identifier stored in the tag. Then, it contacts an uGASP server [11-12]. After consulting an internal database, this server indicates to the mobile phone what must be done: display a text, an image, or play an audio comment. Thus the author of the painting is able to communicate with the visitor.

Via Mineralia is a pervasive serious game which goal is to enrich the visit of a Freiberg museum [13]. In this game, the visitor uses a PDA equipped with an RFID reader. RFID tags (holding a unique identifier) are dispatched in the showcases which the museum wants to emphasize. When the PDA scans a tag, it sends an HTTP request (with tag's identifier) to a web server. To do so, the PDA uses a Wi-Fi network which covers the whole museum. The server answers to the PDA with multimedia information. The PDA displays them in a navigator.

Touchatag company (formerly Tikitag) sells NFC readers which can be connected to Windows or Mac-OS personal computers, and NFC tags dedicated to Touchatag [14]. A customer can then connect to http://www.touchatag.com web site, and define the reaction to be associated to the reading of one tag. When the NFC reader reads a tag, it contacts the Touchatag application which runs permanently on customer's personal computer. Then, via the Internet network to which the computer is connected, this application contacts a Touchatag service called *Application Correlation Service* (ACS). Touchatag application gives tag's identifier to ACS. Then, ACS queries Touchatag database to find reaction associated to the reading of this tag. It sends back this information to Touchatag application. The touchatag application reacts in the appropriate way. For instance, let's assume that the customer has specified the following action on Touchatag web site: when tag *r* with identifier *i* is put on the reader, customer wants her browser to access to Uniform Resource Locator (URL) of a web site *w*. Then, when customer puts tag *r* on the reader, Touchatag application contacts ACS with identifier *i*. ACS replies with URL of *w*. Then, Touchatag application opens a browser with this URL *w*.

Skylanders is a video game developed by *Activision* company [15]. It requires the use of plastic Figures. These Figures contain an NFC tag. When a player puts her Figure on top of a "Portal of Power" (actually, an NFC tag reader), the video game reads the identifier stored in the NFC tag. Then, the game contacts a server to get the information concerning the character which must be displayed: The Figure becomes alive on the screen. Notice that, according to [16], information is also stored inside the tag: Thus the game can work without using a global network to contact a server. This means that Skylanders not only uses a centralized architectural pattern, but also a distributed one.

Based on all of these examples, next section analyzes centralized architectural pattern.

3.2. Analysis

Concerning the functional attribute, any transponder which wants to modify the avatar of a tag does so by sending a modification message to the server. Thus the server is always aware of the last update done on any avatar. As a reader always queries the central database to know the avatar of a tag, it is not possible that the read value is stale. Moreover, knowing a tag identifier, a reader is able to query the server to know the avatar associated to this identifier: the reader is able to know the avatar of a remote tag. As a mobile device queries the server to know the avatar of a remote tag, it is sure that the returned value is not stale.

Concerning the scalability attribute, the maximum number of tags which can be handled by this architectural pattern is limited by the number of avatars which can be stored in the central database. Let *s* be the average size in bytes of an avatar. Let *Scentral* be the maximum size in bytes of the database. We neglect the storage of the link between tag identifiers and avatars in the database. Moreover, we neglect the overhead due to the storage of data in the database. Then, the maximum number of tags is bounded by *Scentral/s*. About sensitivity to the number of simultaneous reads, this architectural pattern is restrained by its centralized nature. The server holding the ONS lookup service may become a bottleneck. Moreover, the different servers of avatars may not return avatar values fast enough. Of course, it is possible to increase the number of servers. But that makes the hardware architecture more complex and more costly (from an installation and a management point of view). Thus this architectural pattern may not be applicable for some applications.

Concerning the cost attribute, the reader must always be in contact with the server holding the ONS lookup service and the servers of avatars: a global network is required. On the other hand, this architectural pattern only needs to read an identifier on the tag. And this identifier can be stored in ROM as it is never modified: no RAM is required on the tags. When a new tag is introduced in the system, three operations are required: (i) the tag is linked to the physical entity; (ii) the avatar of this entity is initialized in the central database; and (iii) a link between the tag identifier and this avatar is created into the central database. When all of the tags have to be reinitialized, a program is run on the server hosting the central database. It sets each avatar to its new value.

This section has analyzed centralized architectural pattern according to the attributes presented in section 2. This architectural pattern fulfills all aspects of functionality attribute. But this is achieved with the operational cost of a global network. Another disadvantage is a high sensitivity to the number of simultaneous read operations.

Next section analyzes semi-distributed architectural pattern which compensates the requirement for a global computer network and reduces sensitivity to the number of simultaneous reads.

4. SEMI-DISTRIBUTED ARCHITECTURAL PATTERN

In semi-distributed architectural pattern, mobile RFID-enabled devices (PDAs, mobile phones, etc.) are periodically synchronized with a central database holding all of the avatars (see Figure 2). Then, human operators carry the mobile devices near the entities to which the tags are associated. When a device comes close to an entity, the device reads the identifier of the entity's tag. By querying its local copy of the central database, the device is able to find the avatar of this entity. Any modification of an avatar is done on the local copy. It is propagated to the central database at the next synchronization.

Next section presents an example of this architectural pattern.

4.1. Example

The unique example of use of such architectural pattern is Paris trees management application [5]. Each of the ninety-five thousand trees of Paris avenues is equipped with an RFID tag. Each gardener synchronizes her tablet PC with the central database before a new day of work. During her day of work, whenever a gardener does something to a tree, she identifies the tree thanks to its RFID tag: Her tablet PC modifies the avatar in the local database. Then, in the evening, she synchronizes her tablet PC with the central database. Thus she uploads her database updates and downloads updates from other gardeners.

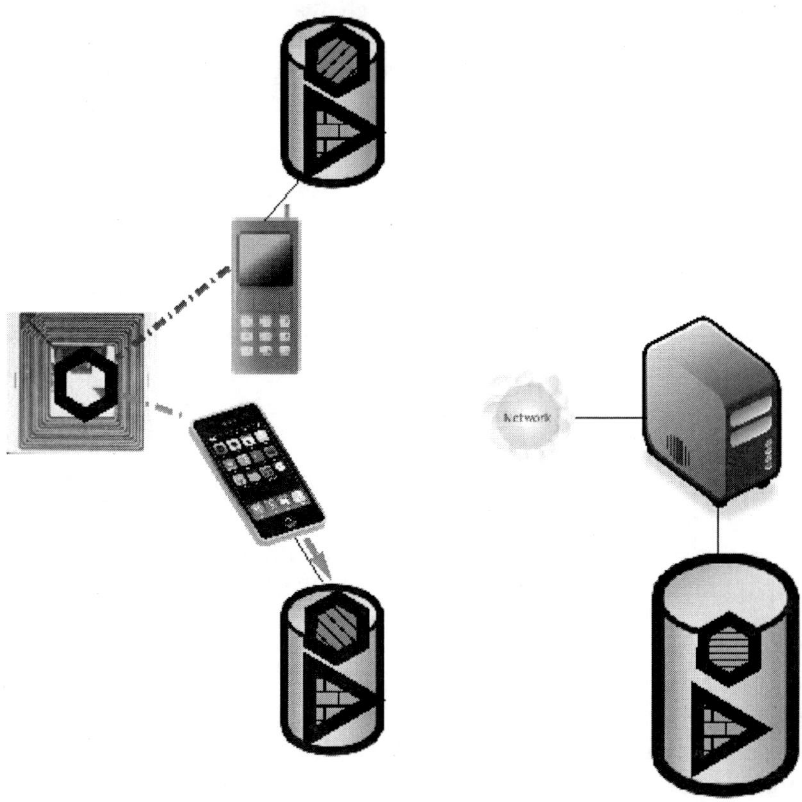

Figure 2. Semi-distributed architectural pattern

Now, let's analyze the semi-distributed architectural pattern.

4.2. Analysis

Concerning the functional attribute, the avatar of a read tag may be stale. Suppose users *U1* and *U2*synchronize their mobile device with the central database. Then *U1* modifies avatar of tag *r*. Thus she modifies her local copy of the database. When *U2* comes to tag *r*, as her device reads its local copy of the database, the returned value of the avatar is the value before *U1*'s modification: the read value is stale. Notice we can limit this issue by assigning sets of entities to each mobile device. For instance, in the case of the Paris trees management application, a supervisor can assign a set of trees to be taken care of during the day, to each of the gardeners. If all of these sets are apart, this issue cannot be observed anymore. About remote tags, by querying its local database, the device is able to read the avatar of a remote tag, even though there is no global network. But the read value can be stale. It will be again correct only when all of the mobile devices have synchronized themselves with the central database.

Concerning the scalability attribute, the maximum number of tags which can be handled by this architectural pattern is limited by the number of avatars which can be stored in the central database and in the local copy of this database. Let *Slocal* be the maximum size in bytes of the local database. The maximum number of tags is bounded by *min(Scentral/s,Slocal/s)*, which is likely to be *Slocal/s* as mobile devices do not have as much memory as servers. Notice that this bound can be increased to*Scentral/s* by assigning to each mobile only a subset of the central database. For instance, in the case of Paris trees management application, the mobile device of a gardener could receive only the avatars of the trees she will take care of during the day. About sensitivity to the number of simultaneous reads, this architectural pattern is not as sensitive as centralized architectural pattern. It does not need to query a server upon each RFID tag read. Nevertheless all of the readers must periodically synchronize themselves with the central database. As the synchronization time is proportional to the number of readers, it may reach unbearable values. This issue can be tackled by limiting the number of avatars copied on the local devices, thus reducing the volume of data transferred between each device and the central database.

Concerning the cost attribute, the mobile RFID-enabled devices only need an access to the server hosting the central database during synchronization phase. At that moment, devices are probably near the central database: A Wi-Fi network may be used. Otherwise it is the local database which is queried. Thus no global communication network is required around the working area. Moreover, as in centralized architectural pattern, there is no need for RAM on the tags. When a new tag is inserted in the system, the procedure to be applied is the same as in the centralized architectural pattern. When all of the tags have to be reinitialized, a program is run on the server hosting the central database. It sets each avatar to its new value. However, the reinitialization of the tags will be effective only when all mobile devices will get synchronized with the central database.

This section has analyzed semi-distributed architectural pattern according to the attributes presented in section 2. This architectural pattern does not require any global network and has a medium sensitivity to the number of simultaneous tag reads. Nevertheless it faces a functional issue concerning the staleness of avatar read on a local (or remote) tag.

Next section analyzes the distributed architectural pattern which tackles the sensitivity and staleness issues.

5. DISTRIBUTED ARCHITECTURAL PATTERN

In distributed architectural pattern, the avatar of an RFID tag is stored inside the RAM of the tag (seeFigure 3). Whenever a user is in contact with a tag, the reader works with the part of the RAM containing the avatar.

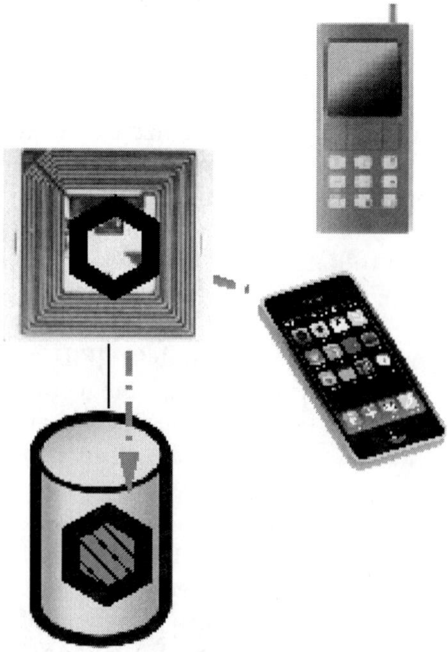

Figure 3. Distributed architectural pattern

Next section gives examples of this architectural pattern.

5.1. Examples

Nokia 6131 NFC phones are sold with three NFC tags. Each one triggers a different function on the telephone: One activates alarm function; another one plays a given music on the phone; the last one displays an NFC tutorial. To do so, the telephone reads the contents of the tag, this contents being coded as a Uniform Resource Identifier (URI) according to the NFC Forum's specifications of *Smarts Posters* [5,17-18]. When the phone is programmed to understand tags' contents formatted according to these specifications, these URIs can be used to tell the telephone to accomplish a given function like send an SMS, call a certain number, open a given web page, etc.

In fact, it is thanks to this Smart Posters specification that any NFC phone can exploit Touchatag tags mentioned in section 3.1. Indeed, these tags contain not only an identifier used by Touchatag application, but also an URI. This URI is the URL of a Touchatag web server with a parameter containing the identifier of the tag. Thus when a user touches a Touchatag tag with her mobile phone, the phone reads the URL and then opens a browser with this URL. Touchatag web server is then contacted, via 3G or Wi-Fi, with the identifier *i*. Then, web server contacts ACS (see section 3.1) with*i*. In the case where *i* is associated with a web site *w*, URL of *w* is sent back to Touchatag web server. This server returns an html page containing a redirection towards *w*. Finally, the browser displays *w*. Notice that, in the case of a Touchatag tag read by a mobile phone, phone uses distributed architectural pattern to determine the Touchatag web server to contact; but, the Touchatag web server uses centralized architectural pattern to translate the tag identifier into an action.

Once again, it is the Smart Posters specification which is used by *Connecthing* company to bring intelligence to mailboxes [19]. When a user scans a mailbox equipped with an NFC tag, her phone reads the URL stored on the tag (which contains an identifier corresponding to the physical location of the mailbox) and opens a browser to access this URL. This web page displays location of nearby mailboxes, the time at which postman takes the mail, etc.

Navigo, the Paris public transportation pass, is an example of an industrial application based on this architectural pattern, which does not use Smart Posters specification [20]. The 4.5 million Navigo pass users do not have an NFC reader. They are only given a pass which contains an NFC tag. With a vending machine, each user initializes her tag with the rights she buys to use the public transportation. Whenever she wants to use a public transportation, she presents her pass in front of an NFC reader. Locally, the reader checks the rights stored in the tag's RAM and opens the gate, if the access is granted.

Ubi-Check is an academic application example of distributed architectural pattern [21]. An RFID tag is attached to each of a traveler's items. At the beginning of their travel, each tag is initialized with a value specific to the traveler. All of these RFID tags are read after special points (e.g. after an airport security control). If an inconsistency is found among the read values, it means that, at some point, the traveler exchanged one of her items with the item of another traveler. An alarm is thus triggered to warn the traveler that one of her items is missing.

[22] proposes an academic system based on digital pheromones to find objects lost in a house. To do so, floor of the house is covered with RFID tags. An RFID reader is coupled with each house object. When user moves an object from point *A* to point *B*, the RFID reader associated with the object behaves like an ant which sets pheromones on the path it takes: The reader writes a digital pheromone (made of object identifier and timestamp of transit) in the RAM of each tag over which it goes. Notice that, like a natural pheromone which evaporates with time, whenever a reader finds no more room in the RAM of a tag (there are too many pheromones stored inside), the reader deletes the oldest pheromone from the tag. In case an object is lost, user takes a dedicated RFID reader and wanders around the house until she finds the digital pheromones of the object. Once she has located it, she follows the pheromone trace until the place where the object was left.

Roboswarn is an (academic) application to position robots (equipped with NFC readers) in a physical space to accomplish a certain task [23]. NFC tags are dispatched in dedicated places of a room (for instance, near a hospital bed which these robots will have to push so that a cleaning robot can accomplish its task). Each tag is initialized with location of other tags in the room and the timestamp of last cleaning. When robots enter the room, they look for an NFC tag. As soon as one robot finds one, it reads the position of other tags and transmits them to other robots. The other robots go to the other tags. If timestamp of last cleaning is too old, robots push the hospital bed and then write new timestamp of cleaning. Otherwise, robots do nothing.

SALTO Systems company is selling locks for electronic doors. The keys are NFC tags. To facilitate the management of all locks and tags, this company has developed SALTO Virtual Network (SVN) [24]. Thanks to this system, Heathrow airport operator is able to manage 1000 standard electronic locks (NFC-controlled) and 37 hot spots. These spots are special locks connected to a global computer network. They can: 1) unlock an entry access on the whole site, 2) initialize an NFC key with the right to open given locks during the day, 3) blacklist some NFC keys, 4) recover data collected by the key during the working day of its user. Indeed, each time a person unlocks an electronic lock with her NFC key, the lock reads data stored on tag to check user permissions and the list of blacklisted tags. But, the electronic lock also writes information like, for instance, the low charge of the battery powering the lock. Thus thanks to SVN, even though standard locks do not have access to a global computer network, they can receive information (e.g. list of blacklisted cards) and send information (e.g. low charge of battery): Standard locks communicate thanks to the network made of the users of the keys/tags.

Based on all of these examples, next section analyzes distributed architectural pattern.

5.2. Analysis

Concerning the functional attribute, as the avatar is written and read only in the RAM of the tag, there is no staleness issue of locally read tags. However, it is impossible to know the avatar of a remote tag.

Concerning the scalability attribute, there is no limit on the number of tags in the application environment. Moreover, such distributed architectural pattern is not sensitive at all to the number of simultaneous read operations (all of the operations are done locally).

Concerning the cost attribute, the reader does not need any global network to access to the avatar of the RFID tag. On the other hand, RAM is required on each tag. Its size must be at least the size of the avatar. This means that the avatar cannot contain too much information (e.g. MIFARE tags can offer up to 4 Kbytes of RAM, with 3440 bytes of net storage capacity). When a new tag is introduced in the system, only two operations are required: (i) the tag is linked to the physical entity; and (ii) the avatar of this entity is initialized in the RAM. About the reinitialization of the tags, it is application-dependant. Some applications require that a dedicated user goes through all of the tags to reinitialize them. In the case of Navigo pass, users are in charge of bringing their pass to a vending machine. This leads to long waiting lines at the beginning of a month, when users must initialize their rights for this month. This is why Navigo operator carries out experiments where users can initialize their tag using a dedicated NFC reader connected to their personal computer. To avoid reinitialization costs, some RFID-based distributed applications put in place special mechanisms. These mechanisms take into account elapsed time in order to automatically reset data. In Roboswarm application (see section 5.1), there is no need to reset the timestamp to trigger a new cleaning of a room. Each robot is aware of a deterioration level. Thus if the timestamp plus this deterioration level is greater than current time, it means that the room needs some cleaning again. With application for pheromone-based object tracking (see section 5.1), although tags have limited RAM capabilities, there is still no need to have a periodic session initialization which would clean up outdated pheromones. Each pheromone is written on a tag with a timestamp. Thus when the device attached to the roaming object meets a tag, it cleans up pheromones which have a too old timestamp, before writing the dedicated pheromone.

This section has analyzed distributed architectural pattern according to the attributes presented in section 2. This architectural pattern does not require any global computer network. And it is not sensitive to the number of simultaneous read operations. Nevertheless it faces a functional issue: it is not possible to get the avatar of a remote tag.

Next section analyzes RFID-based DSM which tackles this issue.

6. RFID-BASED DISTRIBUTED SHARED MEMORY ARCHITECTURAL PATTERN

RFID-based distributed shared memory (RFID-based DSM) mixes the qualities of semi-distributed architectural pattern and distributed architectural pattern [7]. The avatar of an RFID tag is stored in the RAM of the tag. In addition, each tag and each mobile device of the application environment holds a local copy of all of the avatars (see Figure 4). Moreover they hold a vector clock (see Figure 5). Each element of this vector clock is a number corresponding to the last version of the avatar which the tag or the device has learnt about (this is why [25] gives the name *version number* to this number). When a mobile device comes to a tag and modifies the avatar of the tag, this mobile increments the element of the vector clock (stored on the tag and inside its own memory) corresponding to the avatar of this tag. Whenever a mobile device meets a tag (respectively another device), the device and the tag (respectively the other device) compare their respective view of the avatars, by comparing their vector clocks values. Doing so, each of them learns from the other one the latest news (which they are aware of) about all of the avatars.

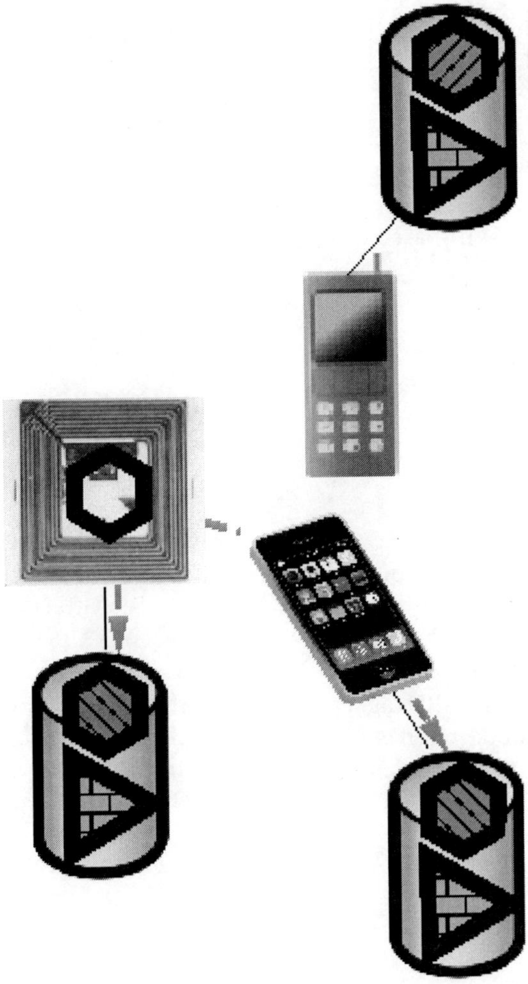

Figure 4. RFID-based distributed shared memory architectural pattern

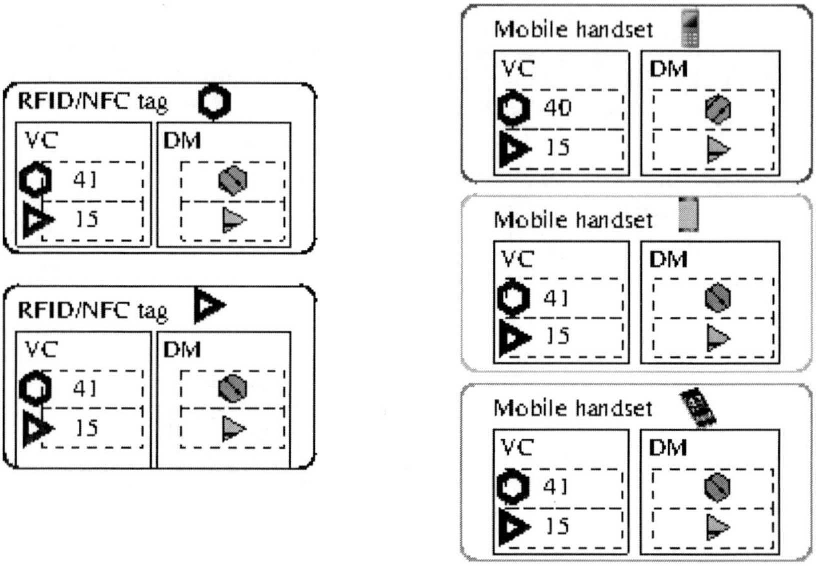

Figure 5. Data of RFID-based distributed shared memory

Next section presents an example of this architectural pattern.

6.1. Example

Plug: Secrets of the museum, an (academic) pervasive game [26] developed in the context of the*PLUG* research project [27], is the unique example of use of RFID-based DSM architectural pattern. In this game, 48 virtual playing cards represent objects of French Museum of Arts and Crafts (*Musée des arts et métiers*). These cards are dealt between 16 NFC tags (1 card per MIFARE tag, each of them being equipped with 1 KB of RAM) and 8 mobile phones (4 cards per Nokia 6131 NFC mobile). The players' goal is to collect cards of the same family on her mobile. To do so, players use their mobile to swap a card with a tag or another mobile.

Next section analyzes RFID-based DSM architectural pattern.

6.2. Analysis

Concerning the functional attribute, whenever a mobile device comes near a tag, there are two possibilities. Either the tag has been already initialized; in that case, as the avatar is stored on the tag, the value read on the tag is the most up-to-date. Or, the tag has not been already initialized; in that case, the first task of the mobile device is to initialize the tag; so that the value read on the tag after this initialization is also the most up-to-date value. Thus there is no staleness issue for avatar of locally read tag. Moreover, a mobile device holds a local copy of all avatars. Thus by querying this local copy, the device is able to answer to queries concerning a remote tag. However this local copy may not be up-to-date: There is a staleness issue for avatar of remotely read tag.

Concerning the scalability attribute, the maximum number of tags is limited by the size of the RAM of the tags. This architectural pattern stores copies of the avatar of all of tags and a vector clock. Let *Stag*be the lowest size of the RAM of the tags present in the environment. Let *L* be the length of an element of the vector clock. Then the maximum number of tags is bounded by $Stag/(s + L)$. Let's compare this bound to the bound of semi-distributed architectural pattern. For the latter pattern, the numerator is expressed in terms of Gigabytes. However it is expressed in terms of Kilobytes in the case of a tag's RAM. The maximum number of tags for RFID-based DSM architectural pattern is at least one million times lower than the maximum number for semi-distributed architectural pattern. About sensitivity to the number of simultaneous reads, RFID-based DSM requires that all mobile devices synchronize with a dedicated machine: RFID-based DSM is as sensitive as semi-distributed architectural pattern.

Concerning the cost attribute, in RFID-based DSM architectural pattern, the reader does not need any global network to access to the avatar of the RFID tag. Nevertheless RAM is required on each tag. Its size must be at least the size of the avatar times the number of tags in the environment (so that a tag can hold a local copy of all avatars). This means that the avatar can contain even less information than in the case of distributed architectural pattern.

When a new tag is introduced in the system, four operations are required: (i) the tag is linked to the physical entity; (ii) the avatar of this entity is created and initialized in *DBinit*, the database used to (re)initialize the local copy on each mobile (*DBinit* is stored on a dedicated machine which can be one of the mobiles); (iii) a link between the tag identifier and this avatar is created in *DBinit*; and (iv) all of the elements of the tag's vector clock are initialized to zero.

When tags must be reinitialized, a program *Pinit* is executed on the machine hosting *DBinit*. This program computes the initial value *VCinit* of the vector clock for this session, so that each element of*VCinit* is greater, thus more recent, than all the vector clock elements in the mobiles. To do so, *Pinit* can use two methods. The first method is twofold: (i) get the vector clocks of all of the mobiles; and (ii) compute the maximum value. This first method does not require additional memory on each tag, but requires additional communication between the mobiles and the dedicated machine. This method works because the vector clock of a tag evolves only when a tag is in contact of a mobile. Thus there is always at least one mobile device which is aware of the values stored in the vector clock of a tag:*Pinit* does not need to be aware of the vector clock values stored on the tags. The second method supposes that each vector clock element is made of two fields: a "session identifier" field and a "tick in this session" field. Thus *Pinit* has only to increase the session number and set all "tick in this session" fields to zero. This second method requires additional memory on each tag, but no additional communication between the mobiles and the machine running *Pinit*. The choice of the method is application dependant. Once one of the two methods has been applied, the dedicated machine synchronizes each mobile device by sending the contents of *DBinit* and *VCinit* to the device. Afterwards, whenever a mobile device is in contact with an uninitialized RFID tag for this session, as the mobile device vector clock is greater than the tag vector clock, the mobile device initializes the tag. In other words, RFID-based DSM architectural pattern takes advantage of the fact that application users will go to tags, to initialize them: This pattern uses the communication network made by application users, instead of using a global computer network.

This section has analyzed RFID-based DSM architectural pattern according to the attributes presented in section 2. This architectural pattern does not require any global computer network. And it does not experience the issue of staleness of an avatar of a read tag. Moreover it is possible to query the avatar of a remote tag. Nevertheless this architectural pattern experiences staleness issues when accessing to avatar of a remote tag. And there is a scalability issue in terms of maximum number of tags which can be handled.

By synthesizing the conclusions observed for the different architectural patterns, the next section provides guidelines for choosing the most adequate pattern for a given application.

7. GUIDELINES FOR CHOOSING THE RIGHT RFID-BASED ARCHITECTURAL PATTERN

Table 1 synthesizes the analysis of the different aspects of the architecture attributes made on all of the architectural patterns. In this table, values which are in italic correspond to aspects which are a limitation for this architectural pattern.

If the application requires the best level for all aspects of functionality attribute, then the centralized architectural pattern must be chosen. It is the only architectural pattern which experiences no issues within the functionality attribute. But this pattern has an operational cost due to the requirement for a global network. And this pattern is highly sensitive to the number of simultaneous reads.

Table 1. Comparison of the RFID-based architectural patterns (italic values signal a limit for this architectural pattern)

	Central.	**Semi-distr.**	**Distributed**	**RDSM**
Staleness of locally read tag	No	*Yes*	No	No
Avatar of remote tag	Yes	Yes	*No*	Yes
Staleness of remote read tag	No	*Yes*	n.a.	*Yes*
Maximum number of tags	$S_{central}/s$	S_{local}/s	Infinite	$S_{tag}/(s+L)$

	Central.	**Semi-distr.**	**Distributed**	**RDSM**
Sensitivity to number of simultaneous reads	*High*	Medium	None	Medium
Network required	*Yes*	No	No	No
RAM required on tag	No	No	*Yes*	*Yes*
Cost of introducing a tag (most costly operation)	Link tag to physical entity	Link tag to physical entity	Link tag to physical entity	Link tag to physical entity
Cost of reinit. Tags (most costly operation)	Reinit. Database	Sync. Database	*Go to all tags*	Sync. database

If one of these last two issues is a problem, the system architect must consider the three other architectural patterns. Semi-distributed architectural pattern must be chosen if RFID tags cannot host RAM. This constraint may be due to cost motivations, but also technical constraints (use of low-frequency tags, see section 2).

If there can be RAM on tags, the maximum number of tags must be determined. If it is compatible with RFID-based DSM architectural pattern limitations, then this pattern should be chosen (as it is the least limited pattern for the functionality attribute). Otherwise the system architect should choose distributed architectural pattern (if there must be no staleness issue for read tags) or semi-distributed architectural pattern (if the cost of reinitializing tags is an important factor). Notice that the mixing of distributed architectural pattern and RFID-based DSM may be an interesting alternative. On each tag, we can store its avatar and the vector clock element corresponding to this avatar. Each mobile device holds a copy of all avatars and a full vector clock. By applying RFID-based DSM procedures, we get a solution for the limited maximum number of tags in RFID-based DSM. And in the same time, we solve distributed architectural pattern limitations (as we can query avatar of remote tags and we reduce the high cost of tag reinitialization).

To illustrate the use of these guidelines, let's consider the choice of the architectural pattern for the RFID-based game *Plug: Secrets of the museum* presented in section 6.1.

Each tag costs about 0.10 euro (respectively 1.50 euro) if it has 0 KB (respectively 1 KB with 752 bytes of net storage capacity, *Stag*=752 bytes) of RAM. The avatar of a tag is the virtual card "contained" in the tag. There is a maximum of 16 cards in the game. Thus the avatar is coded as a byte value (*s*=1 byte). Concerning the vector clock, the project uses the synchronization method requiring a session identifier. To have an ever-increasing value, "session identifier" field stores the initialization time. This time is the difference, measured in milliseconds, between the session initialization time and midnight, January 1st, 1970 UTC. This storage requires 8 bytes per tag. Moreover, each tag holds the "tick in this session" field of each avatar stored in the tag.

This field is coded as a short (*L*=2 bytes). It represents a real-time clock, formatted as the number of seconds since the beginning of the game session (A session lasts less than 2 hours: there is no risk of overflow). This clock is the time known by the tag of the last update of the avatar of another tag. It takes about 20 minutes to attach each of the 16 tags to their correct location, so an average of 75 seconds per tag. Linking the tag and the avatar takes about 5 seconds per tag. Initializing the avatar is done in a few milliseconds by an initialization program. For reinitializing tags, synchronizing all of the 8 mobiles with a dedicated machine takes about 1 minute, that is an average of 4 seconds per tag. Notice that synchronization is based on NFC peer-to-peer communication. The project could have saved synchronization time if it has used Bluetooth®, but it would have used more battery.

If the project is going to use centralized or semi-centralized architectural pattern, it will use a dedicated machine for hosting the central database. This machine will be equipped with a 500 gigabytes disk (*Scentral*=500 GB). In the case of the semi-distributed architectural pattern, the project will use half of the micro-SD memory of each mobile phone to host the local copy of the database (*Slocal*=1 GB). If the project is going to use centralized architectural pattern, each mobile will have to be equipped with a SIM card giving access to a UMTS data plan. This will cost 15 euros per mobile per month. If the project is going to use distributed architectural pattern, it will take 13 minutes to go by all of the 16 tags to reinitialize them, so an average of 49 seconds per tag.

We apply these numeric values to table 1. Table 2 synthesizes the results. In this table, values which are in italic correspond to criteria which are a limitation for this architectural pattern.

Centralized architectural pattern requires a global network which costs 120 euros per month. The museum which hosts the game considers it is too expensive. We have to turn to one of the other architectural patterns. As the game must manage 16 tags and as the RFID-based DSM can handle a maximum of 248 tags (as *s*=1 byte), we can choose this architectural pattern. However, if s had been 250 bytes, this pattern could have handled only 4 tags: It would not have fitted. As there must be no issue about avatar of read tags (the game would not be fun), we would have chosen distributed architectural pattern (or combination of distributed and RFID-based DSM patterns, in order to reduce the costs of reinitializing tags).

8. CONCLUSION AND FUTURE WORK

This chapter studies four RFID-based architectural patterns: centralized, semi-distributed, distributed and RFID-based DSM. It compares them according to nine aspects of three architecture attributes: functionality, scalability and cost. Despite their specific limitations, each architectural pattern fits the requirements of existing applications.

Table 2. Comparison of the RFID-based architectural patterns in the case of the game *Plug: Secrets of the Museum* (italic values signal a limit for this architectural pattern)

	Central.	**Semi-distr.**	**Distributed**	**RDSM**
Maximum number of tags if s=1 byte (s=250 bytes)	500 x 109 (2 x 109)	109 (4 x 106)	Infinite (Infinite)	248 (2)
Cost of computer network (per month)	*120 euros*	0 euro	0 euro	0 euro
Cost of tag (per tag)	0.10 euro	0.10 euro	*1.50 euro*	*1.50 euro*
Cost of introducing a tag (in seconds per tag)	80 s/tag	80 s/tag	80 s/tag	80 s/tag
Cost of reinitializing tags (in seconds per tag)	0 s/tag	4 s/tag	*49 s/tag*	4 s/tag

The chapter proposes guidelines for choosing the RFID-based architectural pattern which will best fit a given application requirements. These guidelines are tested in the context of an RFID-based pervasive game.

Future work concerns the analysis of these architectural patterns with respect to security architecture attribute. Security is a measure of the system's ability to resist unauthorized usage while still providing its services to legitimate users [1]. This future work will determine the influence which the level of resistance to security attacks and the cost of implementing such resistance have on the guidelines provided in this paper. Another attribute we would like to study is the fault-tolerance of the different elements of the system.

NOTES

[1] - . An. architectural. pattern. is. a. description. of. element. and. relation. types. together. with. a. set. of. constraints. on. how. they. may. be. used. [1].

REFERENCES

1. Bass L., Clements P., and Kazman R., Software Architecture in Practice, 2nd Edition. Addison-Wesley Professional, April 2003, ISBN-13: 978-0-321-15495-8.

2. Armenio F., Barthel H., Burstein L., Dietrich P., Duker J., Garrett J., Hogan B., Ryaboy O., Sarma S., Schmidt J., Suen K., Traub K., and Williams J., "The EPCglobal architecture framework," GS1 EPCglobal, Tech. Rep. Version 1.2, September 2007.

3. Gonzalez L., RFID: Stakes for the enterprise! (in French). Afnor Editions, October 2008, ISBN-13 978-2-12-465153-5.

4. Roussos G. and Kostakos V., "RFID in pervasive computing: State-of-the-art and outlook," Pervasive Mob. Comput., vol. 5, no. 1, pp. 110–131, February 2009.

5. ITR Manager.com, "City of Paris is taking care of its trees with RFID tags (in French)," http://www.itrmanager.com/articles/ 59758/ 59758.html (last access in July 2012), December 2006.

6. "Smart Poster Record Type Definition, Technical specification SPR 1.1," NFC Forum, 2006.

7. Simatic M., "RFID-based replicated distributed memory for mobile applications," in Proceedings of the 1st International Conference on

Mobile Computing, Applications, and Services (Mobicase 2009), San Diego, USA. ICST, October 2009.

8. Aspire RFID: OW2 Aspire RFID: an RFID suite for SMEs. Available at http://wiki.aspire.ow2.org (last access in July 2012), 2011.

9. Rashid O., Bamford W., Coulton P., Edwards R., and Scheible J., "PAC-LAN: mixed-reality gaming with RFID-enabled mobile phones," Computers in Entertainment, vol. 4, no. 4, pp. 4–20, October–December 2006.

10. Haberman O., Pellerin R., Gressier-Soudan E. et Haberman U. (2009). RFID painting demonstration. In Natkin S. and Dupire J., éditeurs : Entertainment Computing - ICEC 2009, volume 5709 de Lecture Notes in Computer Science, pages 286-287. Springer Berlin / Heidelberg.

11. Pellerin R., Adgeg G., Delpiano F., Gressier-Soudan E., and Simatic M., "Gasp : a middleware for mobile multiplayer games". http://gasp.ow2.org (last access in July 2012), July 2007.

12. Pellerin R., Delpiano F., Duclos F., Gressier-Soudan E. and Simatic M. (2005) "Gasp : an open source gaming service middleware dedicated to multiplayer games for J2ME based mobile phones". In 7th Int. Conference on Computer Games CGAMES'05 Proceedings, pages 75-82.

13. Heumer G., Gommlich F., Jung B. and Müller A. (2007). Via Mineralia - a pervasive museum exploration game. In Proc. of Pergames 2007, Salzburg, AT.

14. Touchatag: Using the Advanced HTTP Application. Available at http://www.touchatag.com/developer/docs/applications/advanced-HTTP (last access in July 2012), 2010

15. Activision: Skylanders : Spyro's adventure. Available at http:// www.skylanders.com/) (last access in July 2012), 2011.

16. Planck S.: Kids going crazy for Activision Skylanders NFC game. Available at http://www.nfcrumors.com/01-17-2012/kids-crazy-activision-skylanders-nfc/) (last access in July 2012), January 2012.

17. NFC Forum. NFC Data Exchange Format (NDEF) - Technical Specification 1.0. Rapport technique, NFC Forum.

18. NFC Forum (2006c). URI Record Type De?nition - Technical Specification 1.0. Rapport technique, NFC Forum.

19. Connecthings : 08/12/2011 - BAL intelligente @ mobulles Paris @ Lab Postal 2011 (Intelligent mailbox, in French). http://www.connecthings.com/node/123 (last access in July 2012), December 2011.

20. Levallois-Barth C., "Navigo: simplification or absolute traceability". FYP éditions, May 2009, ch. 5 – Security and data protection, in Evolution of digital cultures (in French), pp. 173–181, ISBN-13 978-2-91-657113-3.

21. Couderc P. and Banâtre M., "Beyond RFID: The Ubiquitous Near-Field Distributed Memory," ERCIM news, no. 76, pp. 35–36, January 2009.

22. Mamei M. and Zambonelli F. (2007). Pervasive pheromone-based interaction with RFID tags. ACM Trans. Auton. Adapt. Syst., 2(2):4.

23. Zecca G., Couderc, P., Banâtre M. et Beraldi R. (2009). « Swarm robot synchronization using RFID tags. In Pervasive Computing and Communications," 2009. PerCom 2009. IEEE International Conference on, pages 1-4.

24. SALTO Systems: SALTO Networked Locking System - SVN. Available at http://www.saltosystems.com/index.php?option=com_content&task=view&id=62&Itemid=57) (last access in July 2012), 2007.

25. Murphy A. L. and Picco G., "Using LIME to Support Replication for Availability in Mobile Ad Hoc Networks," in Proceedings of the 8th International Conference on Coordination Models and Languages (COORD06), vol. 4038, Bologna, Italy. Springer Lecture Notes on Computer Science, June 2006, pp. 194–211.

26. Simatic M., Astic I., Aunis C., Gentes A., Guyot-Mbodji A., Jutant C., and Zaza E., "'Plug: Secrets of the Museum': A pervasive game taking place in a museum," in Entertainment Computing - ICEC 2009, Eighth International Conference, Paris, France, September 3-5, 2009, Proceedings, ser. Lecture Notes in Computer Science. Springer, September 2009, pp. 302–303.

27. "PLUG: Play Ubiquitous Games and play more," http://cedric.cnam.fr/PLUG/ (last access in July 2012), August 2009.

4

IMPLEMENTATION OF A COUNTERMEASURE TO RELAY ATTACKS FOR CONTACTLESS HF SYSTEMS

Pierre-Henri Thevenon[1] and Olivier Savry[1]

[1] Léti, Minatec, CEA Grenoble, France

1. INTRODUCTION

Nowadays, HF contactless technologies following the ISO 14443 standard are extensively used worldwide. Critical applications like access control or payment require high security guarantees. However, contactless channels are less secure and offer more opportunities for any kind of intrusion than other ways of communication; e.g. eavesdropping and contactless card activation using false reader [1, 2, 3, 11]. Among the attacks on the physical layer, relay attack is the most dangerous because of its simplicity, its impact and its insensitivity to cryptographic protections. It consists in setting up an unauthorized communication between two devices out of their operating range [4, 6]. OnFigure 1, two attackers are able to create a link between the reader and the contactless card without the agreement of the owner. A relay is composed of two elements: a first one close to the reader and called proxy, a second one close to the card and called mole. These two elements communicate together by a wired or a wireless link

A relay attack is thoroughly transparent for current contactless systems and cryptographic protocols. A possible countermeasure is the distance bounding protocol which can add an upper bound for the distance between the two communicating devices.

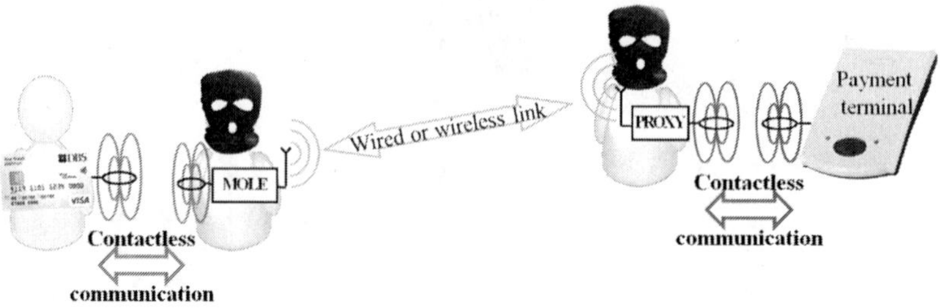

Figure 1. Relay scenario in a queue

In this chapter, we will first assess the potential of relay attacks by implementing them and by keeping in mind the concern of introducing a delay as low as possible. Indeed, this time remains the only detectable feature of such an attack and the existing countermeasures rely on its accurate assessment.

The delay constraint guides us towards the development of three kinds of relay: a wired passive relay, a relay based on a wireless super-heterodyne system and a wireless relay with a complete demodulation of the signal. Our experimental results show that those cheap devices introduce really low delay from 300 ns to 2 μs jeopardizing the use of current distance bounding protocols. A more adapted solution will then be implemented and addressed in the second part of the document. It modifies the stage of the distance bounding protocol which uses the physical layer to carry out a delay assessment with a correlation in the reader between the received signal and the expected one. Finally, a security analysis will be performed and improvements will be discussed.

2. RELAY ATTACKS

2.1. Related Work

The relay attack is based on the Grand Master Chess problem described by Conway in 1976. The latter shows how a person, who does not know the rules of this game, could win against one of two grand masters by challenging them in a same play. The relay attack is just an extension of this problem applied to the security field. By relaying information between a reader and a card outside the reader field, an attacker can circumvent the authentication protocol. This attack needs two devices: a mole and a proxy. The mole pretends to be the true reader and exchange data with the proxy which pretends to be the true card.

The larger the distance between the different elements is, the more efficient is the relay. Typical maximum distances between the reader and the proxy or between the mole and the card are roughly 50 cm. The distance between the mole and the proxy is not limited; it just depends on the chosen technology [5].

By using a relay, an attacker can transmit requests and answers between an honest reader and an honest card separated by 50 metres [6]. Many communication channels can be used to link the mole and the proxy like GSM, WIFI or Ethernet [8]. The delay, introduced by such a relay is more than 15 μs. At the physical layer, this attack is the most dangerous for many reasons:

- The card is activated and transmits information when it is powered, without the agreement of the victim. Anyone can be a victim because the attacker has just to be close enough to control your card like in a crowd.

- The attack occurs on the physical layer i.e. the relay transmits coded bits without knowledge about the frame significance. The ISO9798 standard presents an authentication protocol to prove that the contactless devices involved in the communication share the same secret key. For eavesdropping or skimming attacks, the use of this kind of protocol limits the risks. For the relays attacks, knowing the key is not necessary. Actually, a relay does not neither modify the information of the frame nor has to know its meaning. It just transmits the data. The encrypted data are transmitted as plain text.

- Contactless standards such as standard ISO14443 impose timing constraints in order to synchronize data sent by many cards at the same time, especially during the anti-collision protocol. However, these constraints are not enforced by the majority of cards [9]. These requirements would complicate the relay attack if they were really applied. Another weakness of the standard is the time delay between the reader request and the card answer. These time delay is not only such long but also expandable by the card and consequently by an attacker.

2.2. Presentation Of Relay Attacks

The delay in current relays is mainly due to the use of components such as microcontrollers or RFID chips. This kind of components is used for the reconstruction of the decoded signals. So, the original signal becomes compatible with other protocols, like Wifi or GSM, used in the wireless communication between the mole and the proxy. All these signal processes lead to the addition of delays in the relay. They can be considerably lowered by the only use of analog components. Attack scenarios with wired relays must then be considered because they can induce very low delays. Moreover, this kind of relays is simple to realize, with few cheap components. Even if they seem to be unlikely, they can be effective in a queue for example or if they are hidden in the environment.

2.2.1. Passive Wired Relay

Fig. 2 depicts a simple design of a relay which introduces a very low delay close to a period of the carrier 13.56 MHz. This relay does not require an amplifier or other active components. The coaxial cable between the two antennas can be longer than 20m. Such a system is very low cost; the attacker needs a piece of PCB, few components for the matching and a coaxial wire. Overall cost is a few dollars at most. We claim that wired relays are the simplest and fastest relays by design and as a consequence, they should challenge the approaches of countermeasures which only parry the largest delays.

2.2.2. Relay Based On A Wireless Super Heterodyne System

This relay, shown on fig. 3, is quite similar to the relay attack developed by Hancke because it is not restricted by a wired link. Contrary to Hancke's relay, our wireless relay does not use digital components like microcontrollers or RFID chips to process the signal. The delay induced by this relay should be shorter. To do so, the reader signal of frequency fc is mixed with another signal of frequency F, generated by a local oscillator. It results a signal of frequency fc+F, easier to amplify and to send further. A PLL is used as a local oscillator to have the same frequency in the modulation and demodulation circuit.

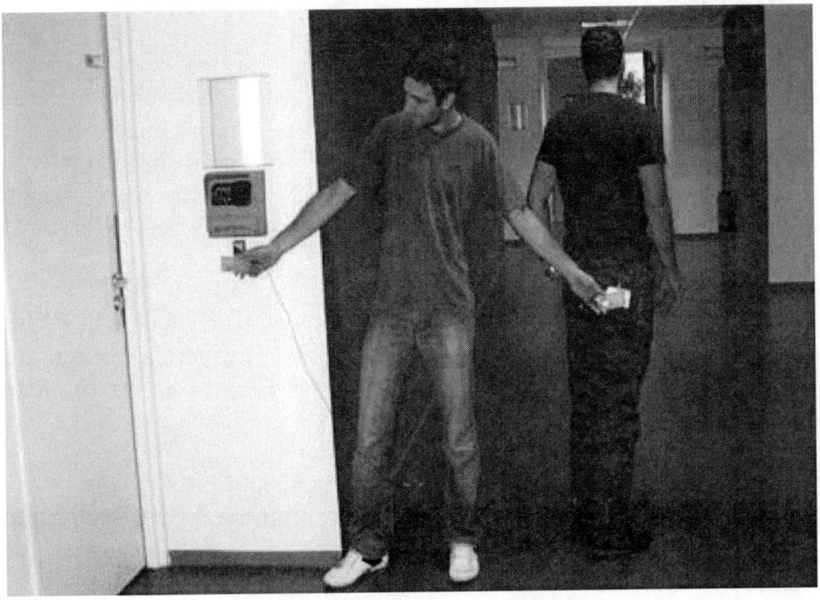

Figure 2. Potential use of a wired relay

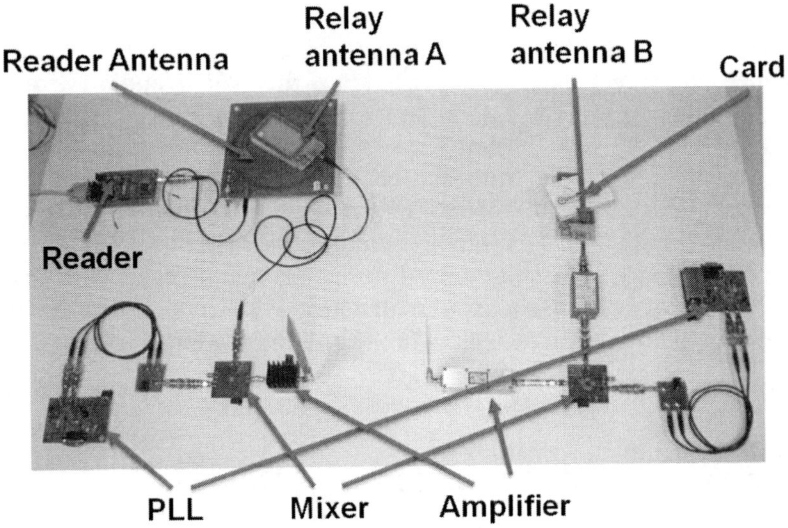

Figure 3. Forward wireless relay

2.2.3. Relay With Demodulation Of The Signal

We have developed a more advanced relay (Fig. 4) close to those realized by Hancke or Kasper.. To realize a relay which demodulates the signal is more complex for an attacker, because it must have a perfect knowledge of the contactless standards. Our system is compliant with the ISO14443-A standard to be compared with literature relays. However, it can be adapted to a different contactless standard such as ISO15693 or ISO14443-B.

The proxy is mainly based on thus developed by Carluccio et al. [7]. This electronic card can be divided into two subsystems: one for demodulation and decoding of the reader signal and one for the load modulation of the card.

The Mole is based on a reader developed in our laboratory. This device has a RF front-end RF which allows amplitude modulation and demodulation. The heart of the mole is a FPGA which separates the phase of emission and reception phase. The proxy signal is processed by the FPGA of the mole; it is coded in modified Miller and modulated in OOK. The HF signal is then amplified and injected in the antenna. The victim's card understands the request of our Mole as a frame from a standard reader and answers by modulating its load. This signal is firstly processed by an analog system and then sampled, demodulated and decoded by the FPGA.

The proxy and the mole communicate together through a wireless system. We have used chips used in the video/audio wireless transmission systems since they allow a sufficient bit rate of 212kbits/s.

The datasheet of video transmission systems provides a theoretical distance operation of 100 metres. In practice, problems of propagation in a building must be taken into account but this distance is sufficient to realize the attack in a shop. Based on experiments realized with the relay, we have obtained a maximum distance of 10 cm between the card and the mole but also between the proxy and the reader.

2.3. Experiments On Introduced Delay

This experiment is performed to measure the introduced delays of the different relays. To do so, a reader sends a fixed sequence through a relay. With an oscilloscope, this sequence is recorded directly on two calibration coils located close to the two relay antennas. This sequence is a signal modulated in amplitude with a subcarrier at 848 kHz. The cross-correlation of the two recorded signals allows the computation of the temporal shift between them. In this experiment, we assume that the delay is the same for the forward and the backward channel so the results are the double of the value which is computed.

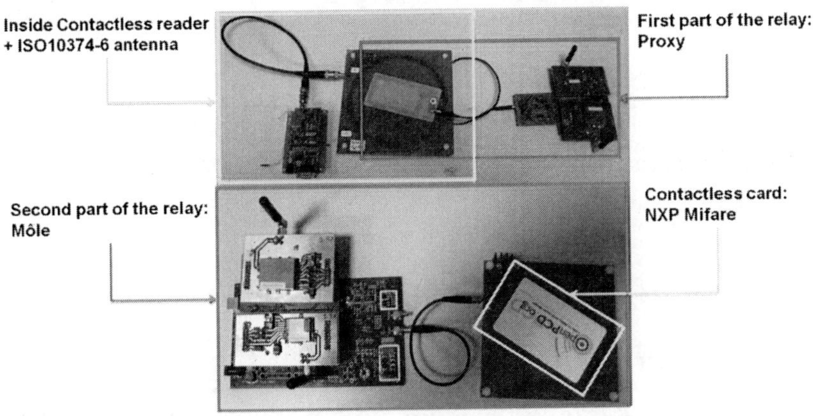

Figure 4. Relay with demodulation of the signal

Fig. 5 gives an overview of the computed delays. Each type of relay is characterized by a temporal distribution. The delay introduced by the relay can be used to detect the presence of a relay. Wireless relays and wired relays have roughly the same delays because the mix of the signals is very fast in the case of the wireless relay. The relay with demodulation introduces a delay 7 times inferior to Hancke's relay.

3. COUNTERMEASURES

In this part, we first describe the main existing relay detection systems and have a critical look to these solutions. Then, a new protocol based on the correlation, is described and implemented on a real contactless system.

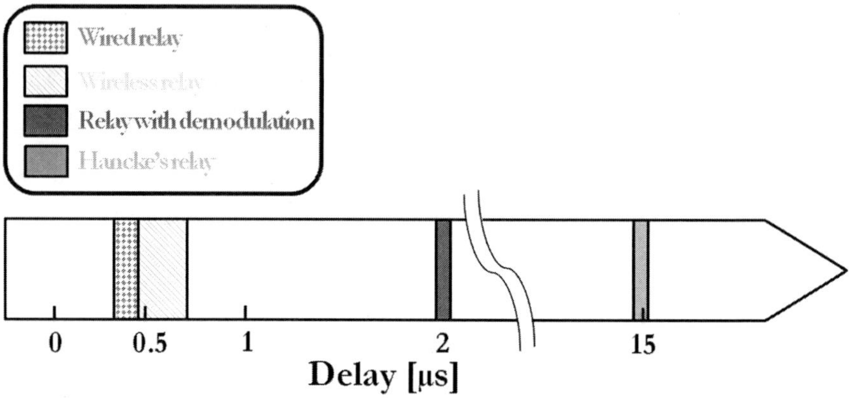

Figure 5. Measured delays

3.1. Existent Countermeasures and Weaknesses

As mentioned before, design of a suitable countermeasure against relay attacks is a veritable challenge. This is partially because cryptography has no effect on it. Currently, there are few methods to detect relay attacks: distance bounding protocols, countermeasures based on timing measurements or physical structures implying the denial of service of the card.

3.1.1. Distance Bounding Protocols

In 2003, Hancke et al. have presented the first distance bounding protocol designed for contactless systems [11]; it is based on Brands and Chaums description [10]. Since then, many others distance bounding protocols have been published to improve the security of the scheme. However, if they have been designed to use the physical layer of the system, they never have been implemented and tested in the HF band.

Distance bounding protocols are used to detect additional delays introduced by a relay during a transaction between two devices. This kind of protocol is often divided into three stages. In such a protocol, the card, named prover, must convince the reader, named verifier, that they are close to each other. During the first stage, the verifier and the prover exchange encrypted sequences, used during the second stage. While, the second stage consists of a timed exchange between the prover and the verifier, in order to verify the card's location. This analysis is made by measuring the time between the request of the verifier and the response of the prover. The last stage is an authentication and verification. The verifier computes and checks the measured times to define the location of the prover and analyses the prover answers to verify its honesty.

The reliability of such protocols depends mainly on the physical layer; the communication channel during the exchange stage affects the accuracy of the propagation time measurements. However, Hancke et al. and recently Rasmussen et al. are the only authors who gave a number of indications related to the protocol implementation at the physical layer level [13]. Other authors claim the merits of their distance bounding protocols such as cost, complexity, reliability but none of them has treated the problem of the protocol implementation for a contactless system.

Such discussions and analysis may be proposed before further works on these protocols.

Distance measurements based on the use of electromagnetic and acoustic waves are used in many applications such as radars. The distance resolution is inversely proportional to the bandwidth; this relation shows one of the weaknesses of distance bounding protocols implemented on a contactless or UWB communication channel:

These two communication channels use electromagnetic waves which have celerity close to the speed of light. In a contactless system, the distance between the verifier and the prover is smaller than 10 cm. Propagation time is then smaller than 300 ps. The first assumption of distance bounding protocol is that the processing time of the signal is assumed to be much smaller than the propagation time of the signal transmitted between the two parts, so smaller than 300 ps.

- HF communication channel: For a contactless system with a bandwidth of 848 kHz, the spatial resolution is around 350 m. Such resolution is too weak to measure a distance between two communicating entities. Moreover, establishing time in HF antennas, processing time for modulation and demodulation take too much time to measure small delays

- UWB communication channel: The bandwidth of a UWB system is equal to 20-25% of its central frequency. The spatial resolution is then close to 1.6 m for a 1GHz UWB system. Such resolution is suitable to detect any kind of relays. However, UWB implementation on an HF contactless system is complex. Hardware constraints are required such as the modification of all RF front-end: add of electrical antennas and specific modulation and demodulation systems.

To summarize, distance bounding protocols are really difficult to implement since the use of the UWB adds cost and complexity. By using HF communication channel, the propagation time remains difficult to isolate because it can be small compared to the processing time. To consider the constraints imposed by the physical layer of HF contactless systems is a priority for developers of algorithms against relay attacks

3.1.2. Solutions Based On Time Measurements

Reid et al. have proposed a solution allowing the measurement of the time duration between the end of the request and the start of the reply [9]. For that, the authors have identified two reference points which represent the state change of the system. In theory, this system can measure average delays of 300 ns; this resolution is 50 times smaller than the delay introduced by Hancke's relay. This counter-measure can be accurate enough to avoid relay attack. However, some problems remain:

- The card does not always reply at the same time;
- No protocol authentication are implemented;
- The signal processing can increase the duration of the delay;
- The attacker can act on the relay to disturb the counter measure.

Munilla et al. have proposed a protocol based on the ISO14443-A standard [10]. In this solution, the reader measures the delay between its request and the card answer. It computes the number of carrier periods between the end of its synchronization bit and the time when the carrier becomes stable after the card response. The authors concluded that their protocol can be used to detect simple relay attacks which induced delays lower than 1 μs. However, this resolution is inefficient against distance fraud attack. Moreover, this countermeasure imposes the modification of standards and of the physical layer. In this solution, the carrier is switched off regularly so the card cannot be powered during this time.

3.1.3. Solution Based On The Denial Of Service

The literature provides some examples of solutions that enable the card's holder to disable their card temporarily [1, 18]. The easier solution is a wallet made of metallic sheets, which acts as a faraday cell. Reference [17] presents physical structures which enable the card's holder to turn off their card by separating the chip and the antenna.

3.2. Our Solution

This part describes a new protocol compliant with contactless standards which authenticates the two communicating parties. A first implementation of this countermeasure on a real contactless system demonstrates its reliability.

3.2.1. The Proposed Scheme

The main objective of this countermeasure is to detect relay attacks by measuring the delay introduced by them by using the correlation method.

The first assumption of our protocol is not based on the propagation time but on the complete delay between a triggering pulse of the reader and the answer of the card received by the reader. This delay is different when a relay is inserted between the reader and the card. For an easier understanding of the solution, we suppose that the forward and backward times induced by the relay are the same to make the explanations easier. In this solution, a recorded sequence is correlated with the same sequence sent by the card. The solution is based on an authentication of the card and the measurement of delay induced by a potential relay, as shown in the fig. 6 and the fig. 7. Our protocol is similar to the distance bounding one; it could be divided into three stages: initialization, time measurement and verification.

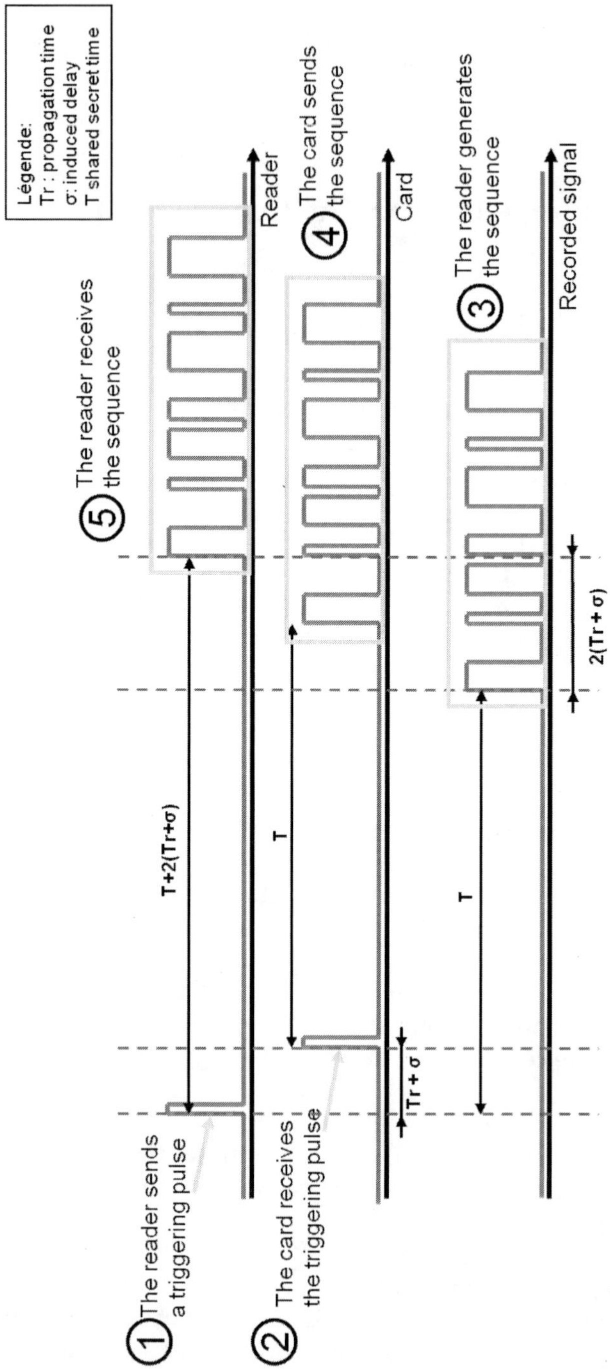

Figure 6. Time measurement stage in the proposed protocol

The first part of our protocol starts by the sending of a nonce from the reader to the card. The reader and the card use any validated symmetric lightweight cryptographic algorithm E, a shared key k and an exchanged random number to calculate T, the waiting time before the sending of the card answer and S, the sequence send by the card and synthesized in the reader. Hence, the computation of T and S by the reader and the card allows the authentication of the card. The first objective of our solution is to detect the relay so there is no mutual authentication in this protocol. However, few modifications of our protocol are possible to have this option.

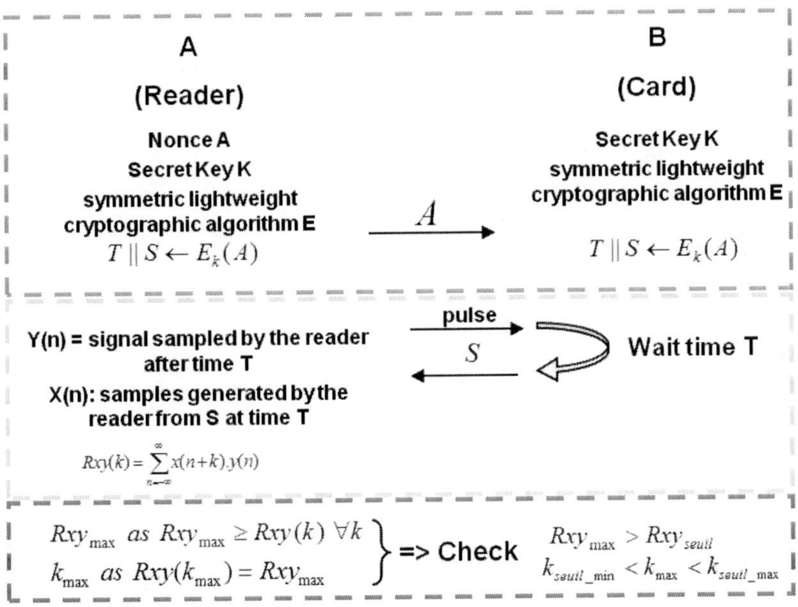

Figure 7. The proposed protocol consists of three stages. The first stage calculates two pseudorandom numbers, using a symmetric cryptographic algorithm and a secret key k and an exchanged nonce. The second stage is time critical as the card has to answer one of the generated pseudorandom sequences, a time T after a synchronization pulse. The third is the verification of the relay presence

After the exchange of the random sequence, the second stage starts (fig. 6). A random number of clock cycles after the end of the request frame, the reader modulates briefly its field to create a synchronization pulse.

This pulse is received by the card with a delay function of the propagation time Tr and the delay induced by the uplink relay σ. It acts as a start point of the protocol for the reader and the card. Once the triggering pulse is received, the card has to send the sequence S after a time duration T measured from the synchronization start by using its load modulation. The time duration between the reader request and the card answer is usually sufficient to send this sequence. The reader received the sequence S with a delay from its sent synchronization. This delay depends strongly on the delays introduced by the uplink and downlink relay. A time T after its synchronization pulse, the reader synthesizes the sequence S as it was sent by the card but without any delay. The received sequence S from the card is sampled by the reader after a time T from the triggering pulse to synchronize the samples Y(n) from the card answer and the sample X(n) from the synthesized sequence of the reader.

During the verification stage, the reader correlates the two recorded sequences X(n) and Y(n) to determine the delay between the two sequences. The index corresponding to the maximum value of the correlation is the number of samples of the delay. This number and the maximum correlation value are used to determinate the presence of a relay in the reader field.

3.2.2. Experiments and Results

This part presents the first results of correlation based on our solution implementation. The solution is implemented on an "open" reader and contactless card that we developed, illustrated in fig. 8.

The objective of our experiment is to demonstrate that the computed delay depends on the relay. We perform our four scenarios: one without relay, the others with the three relays studied previously.

Based on 2000 delay values taken in presence or not of a relay and for different distance between antennas, we compute the distribution of these four cases.

This first implementation of a cryptographic protocol based on the physical layer gives interesting results. The chart on fig. 9 shows three different histograms: one for each implemented relay (the wired relay and the fastest wireless relay) and one in the case without relay. These first results prove the efficiency of our solution since it is able to detect a relay with the help of the maximum delays occurred in a classical contactless system. Only relay designed by us are tested but we assume that delay induced by these relay are close to the theoretical minimal delay induced by the most critical relay. Then, we can claim that our solution can detect the most existed relay attacks.

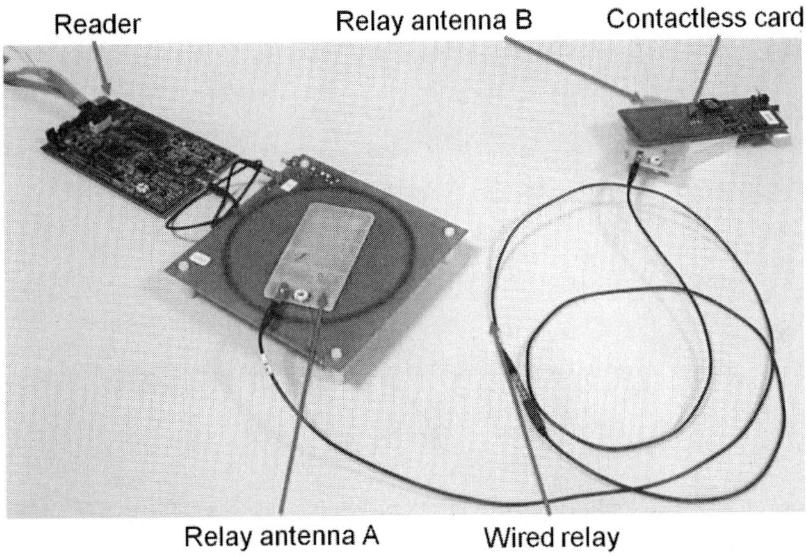

Figure 8. Experimental setup with an open reader and an open card in the presence of a wired relay.

3.3. Discussion

The objective of this discussion is the analysis of the security and the privacy of this solution.

Card cloning and replay attacks with a false card could not be authenticated by the reader and the threat will be detected. In fact, the card must compute two random binary sequences during the first stage of our protocol. The result of this computation is checked during the second stage. A false card could not send the correct sequence at the correct time to the reader because they depend on the knowledge of the secret key k.

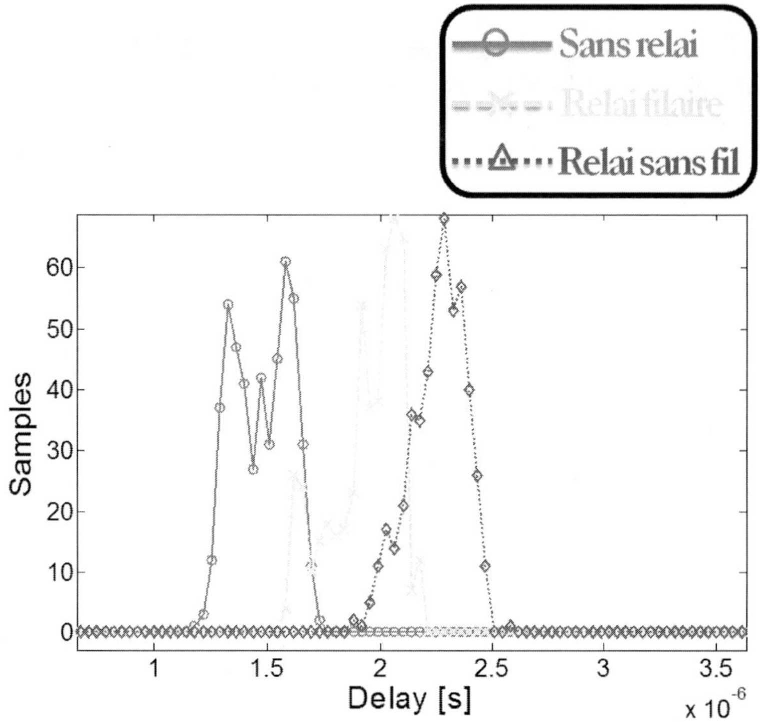

Figure 9. Delay distribution with our solution for each case (with one of the two most critical relays or without relays) for different distance between antennas.

In the case of distance bounding protocols, the security is analyzed by exposing the protocol to three different attacks. Our solution can be exposed to the same attacks to detect possible weaknesses.

3.3.1. Distance Fraud

The scenario of this first attack requires a true reader, named verifier, and a false contactless card, named prover. The prover must convince the verifier they are close to each other when it is outside the communicating range. Firstly, this attack is only theoretical in the domain of contactless systems since no author implements this attack. Thus, the prover authenticates the card during the challenge; a corrupted card will be detected (see above). The detection of distance fraud attacks depends on the delays introduced by a modified card.

3.3.2. Mafia Fraud

In the mafia fraud attack, the attacker does not perform any cryptographic operations based on the security protocol, and only forwards the challenges and the responses between the honest prover and the honest verifier: it is the standard relay attack.To convince the verifier and the prover they are close to each other, the relay can speed up the clock of the carrier to improve the response time of the prover answer [14]. The received signal will be compressed and the correlation value will be weaker so this attack will be detected. In the same way, the relay can not anticipate the synchronization pulse of the card because the pulse position in the time is random. Our protocol is resistant to the mafia fraud.

3.3.3. Terrorist Fraud

This attack is similar to the previous one, the only difference is that the contactless card and the relay cooperate to mislead the reader. This attack is possible if the protocol does not guarantee a link between the authentication part and the timed challenge part. In our case, the answer of the card and the time between the pulse and this answer are deduced by the cryptographic key during the authentication part. Our solution is resistant to terrorist fraud.

3.4. Physical Attacks

The main objective of this article is to prove the reliability of a solution based on the HF physical layer. We assume that the authentication protocol can be improved based on the literature. However, our solution must be resistant to such physical attacks.

3.4.1. Noisy Environment

Distance bounding protocols are usually based on the use of an Ultra Wideband modulation. This modulation is sensitive to noise because its spectral power density is weak. In the case of a noisy channel, the attacker can anticipate the bits sent by the card to reduce the value of the delay measured by the reader. The answer of the card is just one bit; the attacker has a fifty-fifty chance to discover the real value. Then the reader can believe these errors are due to the noisy environment since they are introduced by the attacker. Then the reader concludes that the card is closer than it is and it does not detect the relay (fig. 10).

In our solution, the use of the HF physical layer which is less sensitive to noise and a length of many bits for the sequence S circumvent the anticipation of the sequence by the relay.

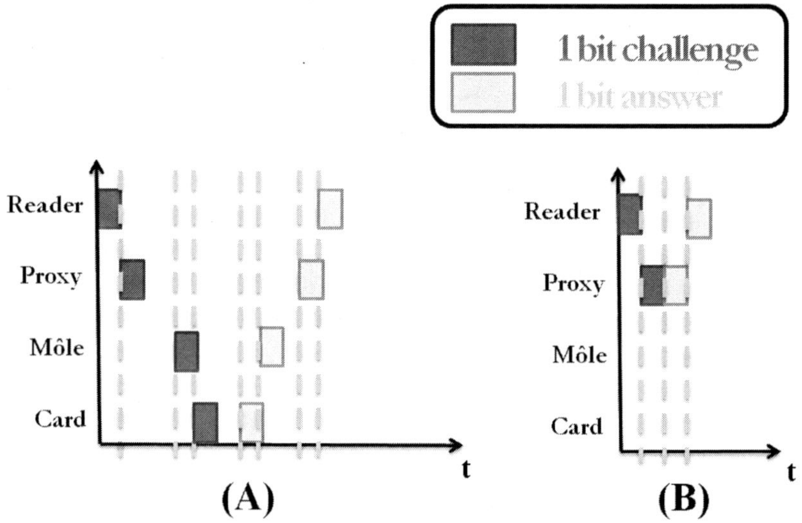

Figure 10. Noisy environment (A) classical case (B) Case with the anticipation of the answer

3.4.2. Timing Attacks

The clock of the card is linked to the carrier frequency of the device which is powered it. This attack, described by Hancke [14], allows an attacker to speed up the clock and then the processes computed by the card to reduce the secret time T of our protocol. Then, the relay transmits the card answer earlier and the relay is not detected (fig. 11). In [16], the authors show that few solutions allow the limitation of the clock increase such as low-pass filters or internal clock. With this kind of solutions implemented on the card, an attacker can absorb 2-3 ns by clock cycle (73.74 ns). To realize such attacks, the attacker has to use a complex relay which demodulates the signal. This kind of systems introduces delays of few µs. Then this attack is not possible if the secret time T between the reader request and the card answer is lower than a determined threshold. This threshold must be inferior to the necessary time to compensate the delay introduced by the processing times of the relay.

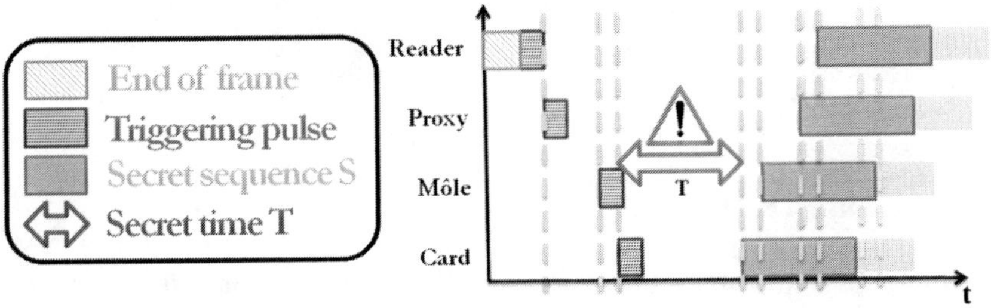

Figure 11. Timing attacks

3.4.3. Anticipation Of The Synchronisation Pulse

The anticipation of the pulse by the relay is a weakness of this kind of protocols because our pulse does not contain a challenge. The relay does not have to wait the pulse and can anticipate and send it earlier. This solution cancels the delay introduced by the forward processing times of the relay (fig. 12). A first solution is to send the pulse just after the end of frame of the reader. Then, the attacker can just cancel the delay introduced by the forward relay. Secondly, our system can use multi level modulation to encrypt the pulse. This modulation can be in amplitude or phase. The value of the secret time T and the secret sequence S can be linked to the value of the modulation level.

Then, this solution limits the anticipation of the pulse since the answer of the card is function of the modulation of the reader.

3.5. Countermeasure Improvement

The accuracy and the reliability of our solution can be enhanced:

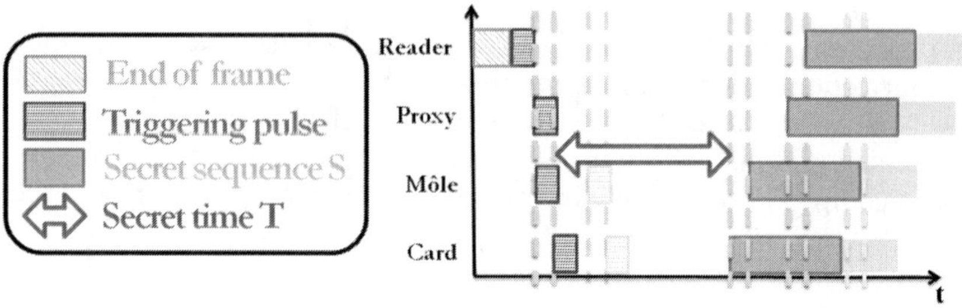

Figure 12. Anticipation of the synchronisation pulse

3.5.1. Pulse Detection

An important improvement concerns the detection of our reference point; the accuracy is mainly due to the triggering pulse identification. It is currently realized using a binary signal, this signal results of the demodulation of the RFID signal. We do not control this demodulation but we suppose that it adds a shifting delay to our total delay. We have to develop a system which can detect a pulse with a fixed delay to reduce delay accuracy significantly. The improvement of the accuracy and the rapidity of the pulse detection can be made by using phase modulation only for the pulse. This solution has been implemented on the previously used contactless reader and a new contactless card able to decode a signal modulated in phase. Our approach, c.f. B.1?, was tested with the new parameters for the pulse emission and reception. The results are described on fig. 13. The delay distribution for the case without relay and the case wired relay show an important improvement. Indeed, the two histograms are significantly different; the introduced delay becomes more important with the presence of the wired relay. This experiment shows that all relay attacks can be detected efficiently using the phase modulation for the synchronization pulse. However, this improvement implies the modification of the existing Radio-Frequency front-end equipment.

3.5.2. M-Sequences

M-sequences present many properties which can improve the accuracy and the sequence generation of our solution. An M-sequence is a pseudo random sequence generated in most cases by linear feedback shift register and is used in many cryptographic applications. Two properties of M-sequences are of interest: randomness and correlation properties. The sequence is composed of pulses with variable width multiple of the minimal period. The autocorrelation of this kind of signals is an approximation of a Kronecker delta function. Such functions present an important peak when there is no delay between the signals is null which is easy to detect in the case of an implementation.

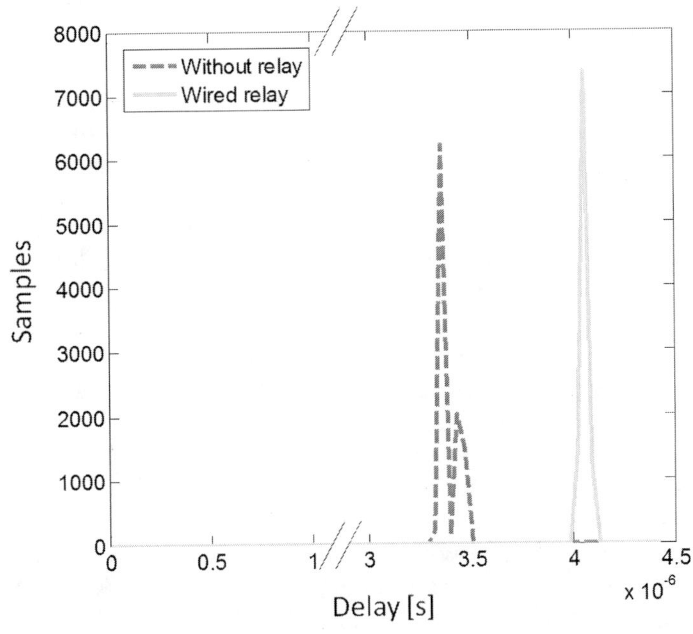

Figure 13. Delay distribution with our solution for the case without relay and in the case with a wired relay for different distance between antennas.

3.5.3. Correlation On Pm (Phase Modulation) Signals

In the case of NFC use in smartphones for critical application, we can suppose that the target (corresponding to the contactless card) uses active mode to answer to the initiator (the reader in our solution).

Then, the target can modulate its signal by varying the instantaneous phase of the carrier signal. The phase modulation can be more complex for implementation but more accurate in terms of correlation. In fact, the signal received and recorded by the initiator must be in phase with the generated one. There are fewer problems with establishing times in antennas because there is no subcarrier, c.f. II.C.2. The obtained accuracy depends on the phase modulation but we can think that we can detect delays close to half of a carrier. Such improvements imply modifications of standards.

4. CONCLUSION

The relay attack is an attack on physical layer which should be seriously considered because it can be easily implemented and used in a lot of applications. Moreover, the increasingly use of NFC technology, especially in phone applications, opens new opportunities for intruders. Nowadays, contactless readers are unable to detect a relay. This attack does not modify the signal, nor disturb the transaction and induce delays close to a few periods of the signal carrier. Additionally, cryptography, which is the best solution for most threats, cannot detect this attack.

The first objective of our work was to realize relay attacks with the shortest delays. Within this chapter, we have presented three different solutions to overcome this problem. Experiment results show that the designed wired relay is the most critical relay in terms of the introduced time delay. Our work shows that with two simple antennas and a wire, an attacker can relay data between a reader and a card with delays close to 300 ns, i.e. 50 times shorter than Hancke's relay attack. Today, no countermeasure is able to detect this kind of relays.

The second objective was to develop a new solution to detect such delays with maximum certainty. This countermeasure uses correlation between two sequences to compute the delay introduced by the relay. This will be used to determine the presence of a relay in the reader's field. For the first time, a solution was implemented on a contactless system and the results are interesting. A contactless system does not require additional hardware resources to use our protocol which allows accuracy close to 300 ns. This solution respects the contactless standards and does not disturb the communication between the reader and the card since the protocol can run during the response time of the card. Apart from the most critical relay, namely wired relay, which is not detected in few rare cases, all kind of relays are detected with our counter-measure. However, we developed another solution that detects all kind of relays attacks by improving the accuracy of our contactless system. However, the latter requires a modification of the RF front end.

REFERENCES

1. Juels A. RFID security and privacy: A research survey, IEEE Journal on Selected Areas in Communications, Vol. 24, Issue 2, p381–394; 2006.

2. Weis S., Sarma S., Rivest R., Engels D. Security and Privacy Aspects of Low-Cost Radio Frequency Identification Systems, proceedings of the International Conference on Security in Pervasive Computing, Vol. 2802, p454-469, SPC 2003; 2003.

3. Hancke G.: Practical attacks on proximity identification systems, IEEE Symposium on Security and Privacy, p328-333; 2006.

4. Hancke G., Mayes K., Markantonakis K. Confidence in Smart Token Proximity: Relay Attacks Revisited, Elsevier Computers & Security, Vol. 28, Issue 7, p615-627; 2009.

5. Lishoy F., Hancke G. P., Mayes K., Markantonakis K. Practical NFC Peer-to-Peer Relay Attack using Mobile Phones, Workshop on RFID Security, RFIDSec'10, 7-9 June 2010, Istanbul, Turkey; 2010.

6. Hancke G. A Practical Relay Attack on ISO 14443 Proximity Cards, Manuscript; 2005.

7. Carluccio D., Kasper T., Paar C. Implementation details of a multipurpose ISO 14443 RFID-tool, Workshop on RFID Security, RFIDsec'06, 12-14 July 2006, Graz, Austria; 2006.

8. Kfir Z., Wool A. Picking virtual pockets using relay attacks on contactless smartcard systems, SecureComm 2005, 5-9 September 2005, Athens, Greece; 2005.

9. Hlavac M., Rosa T. A Note on the Relay Attacks on e-passports: The Case of Czech e-passports, IACR ePrint; 2007.

10. Brands S., Chaum D. Distance Bounding Protocols, Advances in Cryptology, p344–359, Workshop on the Theory and Application of Cryptographic Techniques, EUROCYPT'93, May 23-27, 1993, Lofthus, Norway; 1993.

11. Hancke G. Eavesdropping Attacks on High-Frequency RFID Tokens, Workshop on RFID Security, RFIDSec'08, Budapest, Hungary; 2008.

12. G. Hancke, M. Kuhn, An RFID distance bounding protocol, SecureComm 2005, 5-9 September 2005, Athens, Greece; 2005.

13. Rasmussen K. B., Capkun S. Realization of RF Distance Bounding, 19th USENIX Security Symposium, USENIX'10, 11-13 August 2010, Washington, DC, USA; 2010.

14. Hancke G., Kuhn M. Attacks on Time-of-Flight Distance Bounding Channels: roceedings of the first ACM Conference on Wireless Network Security, WiSec'08, p194–202, 31 March – 2 April 2008, New York, USA; 2008.

15. Munilla J., Ortiz A., Peinado A. Distance Bounding Protocols with void-challenges for RFIDs, Workshop on RFID Security, RFIDsec'06, 12-14 July 2006, Graz, Austria; 2006.

16. Reid J., Gonzalez Neito J., Tang T., Senadji B. Detecting Relay Attacks with Timing Based Protocols, ACM Symposium on Information, Computer and Communications Security, ASIACCS 2007, 20-22 March 2007, Singapore; 2007.

17. Karjoth G., Moskowitz P. Disabling RFID Tags with Visible Confirmation: Clipped Tags Are Silenced, Workshop on Privacy in the Electronic Society, WPES'05, 7 November 2005, Alexandria, VA, USA; 2005.

18. The off switch for "always on" mobile wireless devices, spychips, toll tags, RFID tags and technologies. www.mobilecloak.com (accessed 12 September 2012).

19. DIFRwear's RFID Blocking Products. www.dirfwear.com (accessed 12 September 2012).

5

RFID TEXTILE ANTENNA AND ITS DEVELOPMENT

Lukas Vojtech[1], Robi Dahal[1], Demet Mercan[1] and Marek Neruda[1]

[1] Department of Telecommunication Engineering, FEE,
CTU in Prague, Prague, Czech Republic

1. INTRODUCTION

Textile fabric material has become one of the most important things in life. In early times people used to wear the animal skin to cover their body. The advance form of this is all the clothes we wear today. They protect our body from changing environment conditions and keep us warm. As the technology is increasing day by day, it is influencing every sector. With the increase in wireless technology the electromagnetic radiation also increases. This increased radiation may affect human body severely. Thus with invent of problem, cure was also proposed to make a conductive textile material that could be equally wearable but at the same time work as a filter and does not allow the harmful frequency signal to penetrate into the human body. This completely changed the purpose of fabric material which was previously assumed to be used only for keeping human body warm as now it can be used for protection against the harmful electromagnetic radiation.

Going one step further ahead, we have tried to explore more advantage of textile fabric. With this new invention of conductive textile, we have designed an antenna for RFID (Radio Frequency Identification) applications made out of conductive textile material.

2. RFID BASICS

The RFID uses wireless technology to identify the objects. It consists of RFID tag and a reader. The bi directional communication between the tag and the reader is accomplished by the Radio Frequency (RF) part of the electromagnetic spectrum, to carry information between an RFID tag and reader. There are two types of RFID tag. Passive RFID tags are the ones that does not require any external power supply and works by receiving the signal from reader and retransmit the signal back to reader. Active RFID tag consists of external source in them. These are more complex than passive RFID tags and also give long range communication between tag and reader, when compared with passive tag.

The basic block diagram describes the bi directional communication between the tag and the reader, see Figure 1. The tag antenna in the block diagram receives the RF signal from the reader. This signal is received by the tag antenna, rectified and supplied to the chip to power it up. After the chip is powered up, it now acts as a source and retransmits the signal back to the reader. The reader after receiving the signal sends further to the computer to process the data. The method used to send the signal back to the reader from the tag is called back scattering.

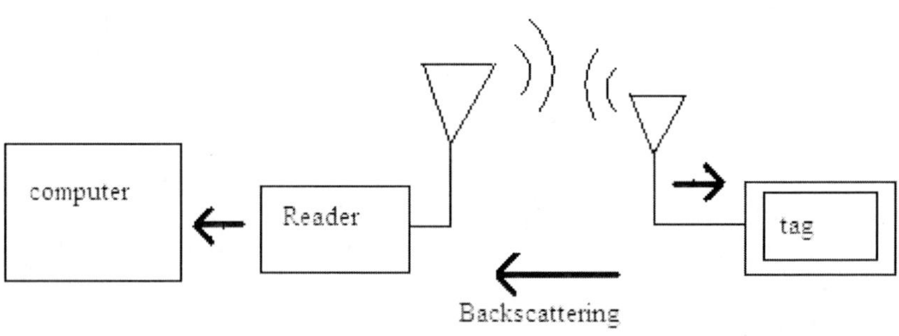

Figure 1. Block diagram of RFID system

2.1. Rfid Transponder

RFID Transponder is basically a radio transmitter and receiver. It mainly consists of two parts, antenna and the integrated circuit (IC). The main function of an antenna is to capture the radiated electromagnetic field by the reader at a definite frequency. The received electromagnetic energy is converted to electrical power and supplied to integrated circuit. The IC chip in the transponder has the capability to store the information to be transmitted to the reader, execute the series of command and also sometimes stores new information sent by the reader [1]. The IC chip mainly consists of a rectifier which rectifies the alternating voltage (AC) received by antenna to the continuous voltage (DC) and supplies to the rest of the circuit in the IC chip.

The IC used for the research is EM4222. This is a read only UHF identification device. The EM4222 is used as a passive chip for UHF transponder. It does not have any internal power supply source. The RF beam is transmitted by the reader. The antenna in the transponder receives the signal, rectifies it and supply the rectified voltage to the chip. The basic block diagram is shown in Figure 2.

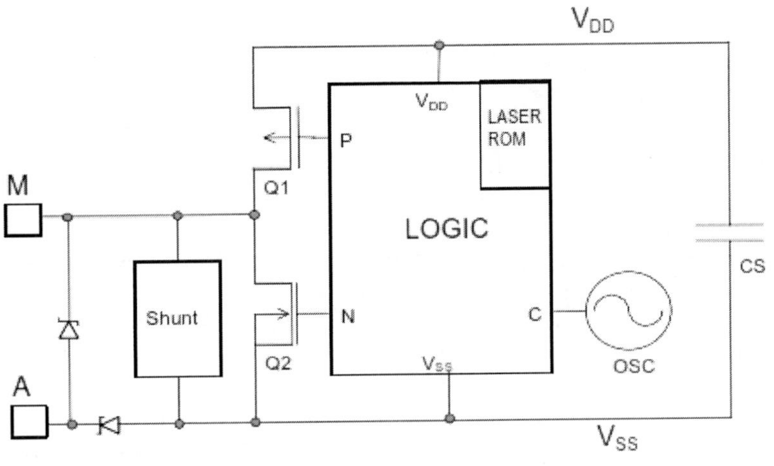

Figure 2. Block diagram UHF transponder EM4222 [**2**]

From the block diagram, it can be seen that the radio beam is received at the terminal *A* in the chip. This signal is rectified to a DC voltage. The shunt regulator is used to limit the input voltage to the logic circuit. It also protects the Schottky diode which is used as a rectifier.

The on chip oscillator in the transponder is used to provide the clock pulse to the logic and also defines the data rate. On chip oscillator present in the transponder oscillates at a frequency of 512 kHz.

If the supply voltage is less than the threshold voltage, the oscillator and the logic cannot function properly and thus the transponder cannot be activate. At this condition, the logic is in reset position. This ensures that the transistor Q2 is off during power up and do not let any false operation to act.

Among the two transistors, Q1 is turned on during power up. Q2 is the modulation transistor which when turned on, loads the antenna with the information from the tag. Q2 is active when the data is to be transmitted from transponder to reader.

In order to have a maximum power transferred from antenna to the chip, the antenna should be designed such that the impedance of the antenna is conjugated matched with that of chip for the given frequency. Generally the chip has capacitive impedance so to have a perfect match the antenna impedance should be inductive in nature.

2.2. Rfid Matching

The tag antenna receives RF energy from the reader. The tag antenna works for a definite resonant frequency. So when the reader transmit RF signal with the desired frequency, the tag receives the signal and supplies to the chip which is attached to it in the transponder. The chip after getting sufficient voltage is able to wake up and hence retransmit the signal at the same frequency to the reader. Thus the purpose of matching an antenna with its load is to insure that maximum power transferred from antenna to chip. To do this, it is needed to have a perfect match between the antenna and the chip. Perfect antenna matching can be achieved by changing the dimension of an antenna, by adding a reactive component or implementing both of them. A mathematical expression can be overviewed as depicted in Figure 3.

The power delivered form antenna to the load or chip is given as [3]:

$$P_l = \frac{R_{ic}}{2[(R_{ant}+R_{ic})^2+(X_{ant}+X_{ic})^2]} V^2{}_{ant} \quad (1)$$

In the above equation it can be seen that the maximum power can be delivered from the antenna to the IC only if *Ric= Rant* and *Xic= - Xant*. Thus it can be observed that the maximum power can be delivered from antenna to load only if they are conjugate matched. This gives one of the favorable conditions for antenna designer as generally the antenna impedance is inductive in nature and the impedance of the chip is capacitive.

Table 1. Electrical characteristics of IC EM4222, VM-VA=2V, TA=25°C, unless otherwise specified [**2**]

Parameter	**Symbol**	**Test conditions**	**Min**	**Type**	**Max**	**Units**
Oscillator frequency Wake up voltage Static Current Consumption Input Series Impedance Input Series Impedance Input Series Impedance	Fosc Vwu I STAT Zin Zin Zin	-40°C to +85°C VM-VA rising VM=1v 869 MHz ; -10dBm 915 MHz ; -10dBm 2.45 GHz ; -10dBm	400 1.0	512 1.4 1 128-j577 132-j553 80-j232	600 1.85	KHz V µA Ω Ω Ω

In this research the antenna is designed to work at 869 MHz. At this frequency the input series impedance of the chip is 128-j577 Ω. Thus the requirement is to have antenna impedance of 128+j577 Ω such that it is complex conjugate matched with the load and maximum power is transferred.

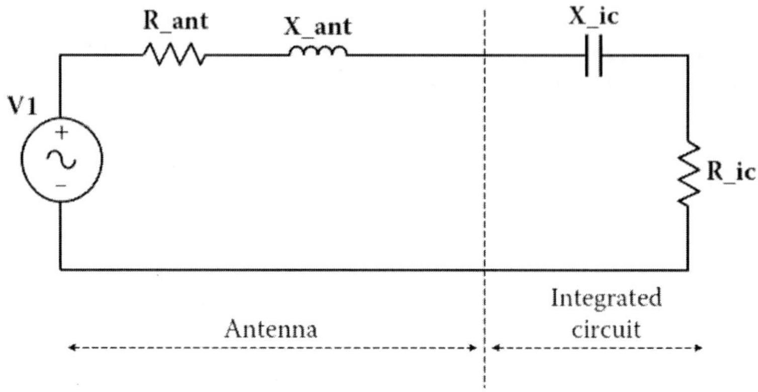

Figure 3. Series model for transponder chip and antenna [**3**]

Conjugate Match Factor (CMF) is the factor which tells how good matching is done between the chip impedance and the antenna impedance. It can be described as the ratio between antenna input power with given chip impedance *Zs* and antenna impedance *Za* assuming *Za* is complex conjugate of *Zs*.

The value of CMF changes between 0 and 1 in linear. To receive maximum power from the reader and retransmit the maximum power to the reader, the antenna impedance should be complex conjugate match and equaled with that of chip.

3. H-SLOT MICROSTRIP PATCH ANTENNA FOR UHF RFID

In this section the passive UHF RFID tag design is discussed. This RFID tag is textile made and involving the human body as the object to be tagged.

The designed antenna layout is an H-shape slot place onto a patch, Figure 4.

The patch with H-slot is placed on a substrate and grounded by a conductive material to decouple from the human body. H-slot is a tuning slot for the required conjugate impedance matching between the microchip and the antenna tag. The maximum size of the antenna is 150 mm x 180 mm and the gain is rather poor around -7 dB due to the bidirectional radiation of the slot. But the maximum gain can be increased by increasing width of the tag antenna. And also the impedance matching is done by tuning the internal slot size.

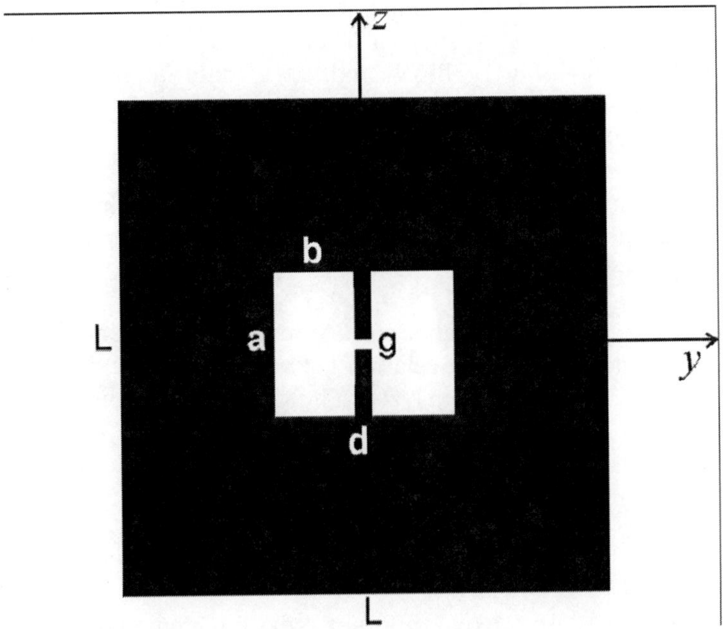

Figure 4. Geometry of nested - slot. The microchip is placed in the central gap of the slot [**4**]

Dielectric material of this patch antenna has a thickness of *h* and it has a longer face of it in the lower part which is placed on the human body through the conductive ground plane. It is an advantage to have longer ground plane because it will avoid the effect of human body radiation.

The radiation is produced by the patch open edge and by the slot. To achieve better radiation performance, width of the antenna can be increased depending of available place for tag. The dimension of the central gap is kept fix by the microchip packing but for tuning the other dimensions of the slot are optimized. The perfect conjugate matching should be done between antenna and microchip to obtain the maximum reading distance.

4. OBSERVATIONS

4.1. Impact On Antenna Performance With Radiating Element Having Different Surface Resistivity.

While working with textile antenna, it is found that the antenna is not working properly as it should work. This is because the antenna which is generally made of very high conductive material has very good radiation efficiency and gain. However the antenna made of conductive textile material has very high surface resistivity and hence lower conductivity. Because of this property of textile material it is difficult to choose the appropriate conductive textile material for desired gain. Also the height of dielectric constant plays a major role in determining the radiation efficiency.

Because of the problem of higher surface resistivity associated with the conductive textile material, a relation between surface resistivity, gain and radiation efficiency is analyzed. For this purpose the microstrip patch antenna is simulated in simulation software IE3D. First the measurement is performed by simulating two different microstrip patch antenna. Both of these antennas are simulated to work at a frequency of 2.45 GHz but with the radiating element having different surface resistivity.

The surface resistivity for two different radiating elements was chosen to be 0.02 Ω/sq and 1.19 Ω/sq. Fleece fabric is used as dielectric material which has a dielectric constant of 1.25. The antennas are simulated for reflection loss less than -40 dB and the result is noted.

It is observed from the simulation result that, though all the other antenna parameter are same, the difference in surface resistivity of the two radiating element affect a lot in their radiation efficiency and gain.

To know the relation between these parameters, a measurement is done for 20 different microstrip antennas keeping other parameters same and only changing the surface resistivity. The result is then plotted in matlab. These measurement results obtained when analyzing different antennas provide valuable information when a conductive textile material is to be used to design an antenna.

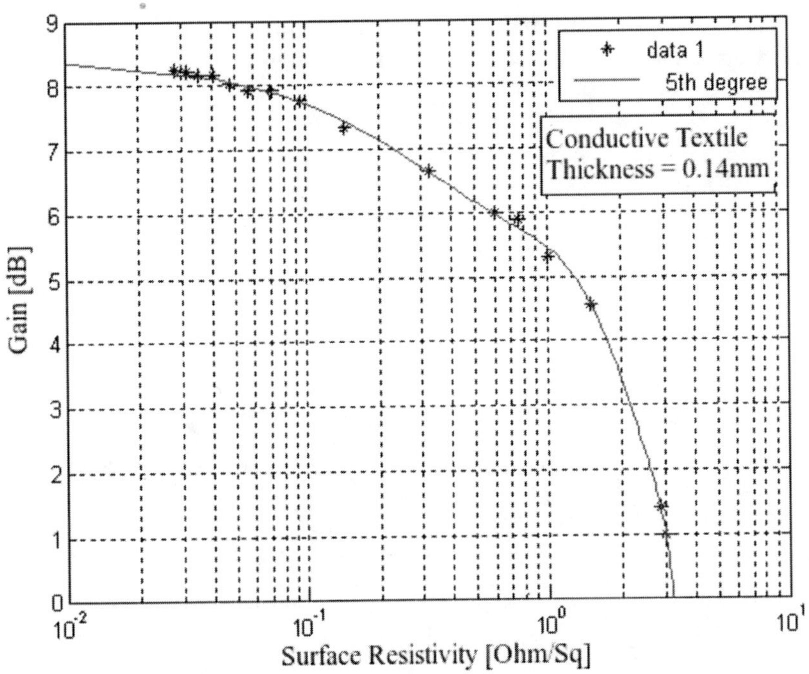

Figure 5. Gain Vs Surface Resistivity Plot

Figure 5 depicts that for very low surface resistivity, the gain is maximum. When the radiating element (antenna) surface resistivity is increased, the gain of the antenna starts to decrease.

Figure 6 illustrates the relation between the conductivity and the radiation efficiency for above mentioned conductive textile material when used as an antenna. Radiation efficiency is a ratio of power radiated by antenna and power input to antenna. If most of the power input to the antenna is radiated, then the antenna is said to have high radiation efficiency. It can be seen from Figure 7 that conductivity is related to radiation efficiency in logarithmic manner. When the conductivity of radiating element is lower, the radiation efficiency is also very small, and increases as the conductivity increases. However radiation efficiency does not increase in linear way, and to achieve the radiation efficiency in higher percentage, the conductivity of the radiating material should be very high.

The entire simulated antenna has reflection (*S11*) less than -35 dB. However the entire antenna does not have same *S11*. This affects the smoothness of the curve obtained in Figure 5 and 6.

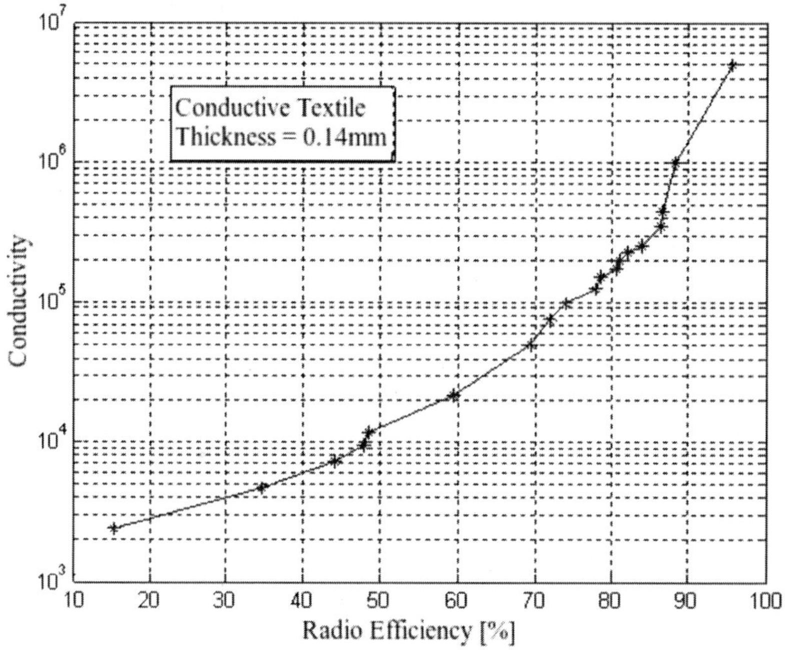

Figure 6. Radiation Efficiency vs. Conductivity Plot

4.2. The Impact On Antenna Performance When The Thickness Of Dielectric Material Is Changed To Different Values.

When the dielectric material thickness is changed, this affects the radiation efficiency. To analyze this effect, three microstrip patch antenna is designed.

Three rectangular patch antennas are designed for different height of dielectric material. On doing simulation, various parameters like reflection, gain, radiation efficiency and antenna efficiency for different patch antenna were observed. The obtained results are shown in Figure 7 and 8.

Table 2. Technical specification for the antenna

Frequency	2.45	GHz
Surface resistivity of radiating element	0.02	Ω/sq
Thickness of radiating element	0.14	mm
Dielectric material	Fleece fabric	-
Dielectric constant	1.25	-
Height of Dielectric material	1, 2, 3	mm

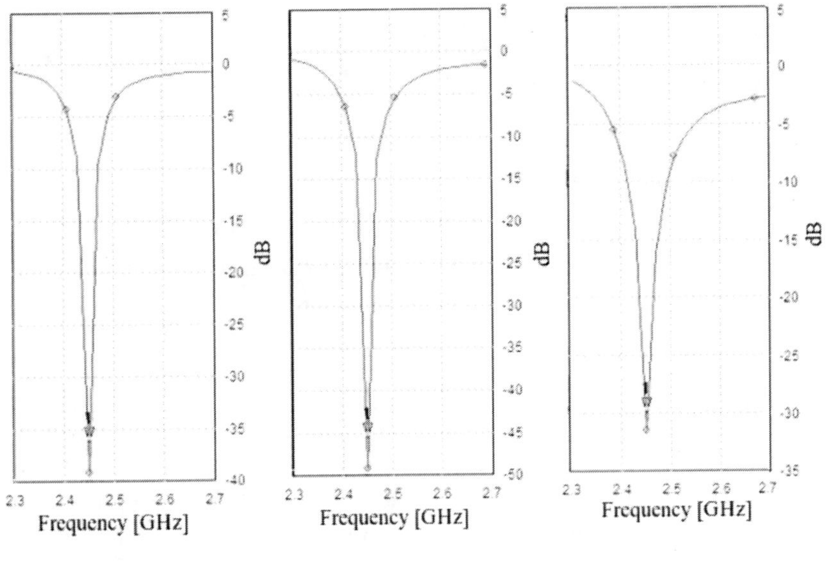

Figure 7. S_{11} for antenna with dielectric thickness 1 mm, 2 mm and 3 mm correspondingly

From above *S11* plot it can be seen that the reflection is less than -30 dB for all three antennas with dielectric thickness 1 mm, 2 mm and 3 mm.

For the same specification of antennas, the radiation efficiency is measured with different height of dielectric material.

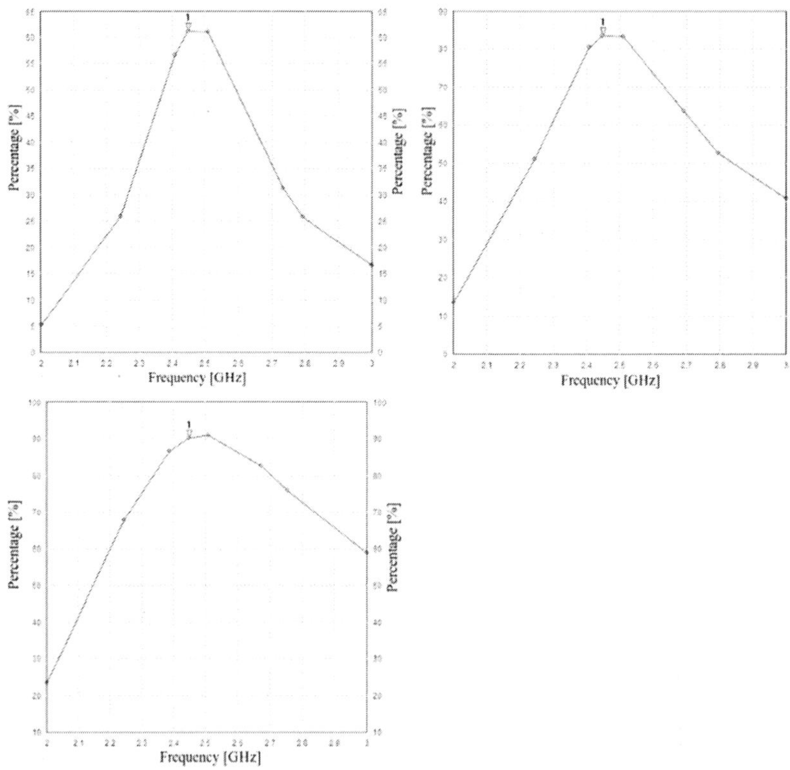

Figure 8. Radiation Efficiency for antennas having thickness 1 mm, 2 mm, 3 mm respectively

The above plot gives the measure of radiation efficiency of the antenna with three different thicknesses. It can be seen that for an antenna working at 2.45 GHz and dielectric thickness of 1 mm, the radiation efficiency is 61.2 %, for dielectric thickness of 2 mm, the radiation efficiency is 83.4 % and for dielectric thickness of 3 mm, the radiation efficiency is 90.2 %.

5. CONDUCTIVE TEXTILE MATERIALS

The fabric that can conduct electricity is called conductive fabric. The conductivity of the fabric depends on how it is manufactured. Conductive fabric can be made in various ways. They can either be produced by metal inter woven fabric during manufacturing or by metal coated fabric [5] also called electro thread. These conductive textiles have wider application in various fields as they are used for shielding human body and some special equipment from external electromagnetic radiation, and also as pressure sensor or flexible heaters, which are made out of easily wearable conductive textile [6].

For a good design of a textile antenna, the conductive fabric should satisfy some of the conditions as given below.

- The electrical resistance of the conductive textile fabric should be small in order to reduce the ohmic losses in the fabric.
- The surface resistivity should be homogeneous over the entire conductive textile fabric i.e. the variation of resistance should be minimum.
- The fabric should be flexible enough to be able to use as a wearable antenna.

The antenna performs better if the conductive textile fulfills the above given characteristics.

Non woven fabric

By the name it can be concluded that non woven fabric is prepared by neither knitting process, nor are woven fabric. Thus the non woven fabric does not go through the initial stage of yarn spinning and also a definite web pattern as that of a woven fabric is not obtained. Non woven fabric manufacturing process is similar to that of paper manufacturing process.

The material used is Cu-Ni with the thickness measured in the lab is 0.14 mm.

These kind of fabric are generally manufactured in three ways namely, drylaid syste, wetlaid system and polymer based system. After the fabric is manufactured, it is then strengthened. There are various ways for strengthening fiber web as by using chemical means by spraying, coating. This can also be achieved by thermal means by blowing air or by ultrasonic impact which partially fuses (connects) the fiber thread. Thus finally the metal layer is coated.

Table 3. Technical specification of Cu-Ni textile material [7]

Description	copper + nickel plated non-woven polyamide fabric
Roll widths	102 cm ± 2 cm
Surface resistivity	Max average 0,02 Ohm/square
Shielding effectiveness	70-90 dB from 50 MHz to 1 GHz
Purpose	conductive fabric for general use
Temperature range	-30 to 90 (degree centigrade)

Woven Fabric

Woven fabric is a construction design for lab use at CTU in Prague. The woven fabric consists of silver nano particles attached to the thread of fiber when being constructed and then woven to form a conductive textile, Figure 9.

Table 4. Technical specification of Betex textile material

Name	Betex
Materials Used	Shiledex (60%), Polyster (40 %)
Number of fiber threads per centimeter	20
Surface Resistivity	1.19Ω /sq.

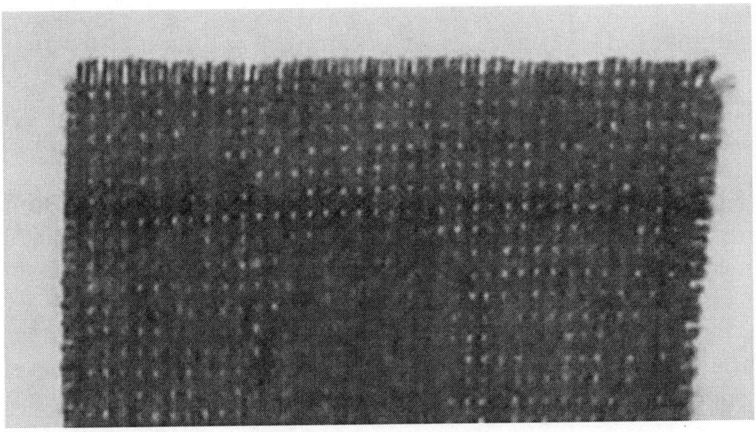

Figure 9. Photo of woven Betex textile material

5.1. Electrical Resistance and Resistivity Of Textile Materials

Electrical conductivity of textile materials is calculated from electrical resistivity as:

$$\sigma = 1/\rho \tag{2}$$

whereσ is electrical conductivity [S/m], ρ is electrical resistivity [Ω m].

Electrical resistivity can be obtained via resistance mesurement [8-10]. We differentiate surface and bulk electrical resistance. Surface resistance is defined as the ratio of a DC voltage *U* to the current *IS*which flows beteen two specific electrodes. The electrodes are placed on the same side of measured material and it is assumed all currents flows only between electrodes and do not penetrate into the bulk of material [8].

Bulk resistance or electrical resistance takes into account all currents flowing in the material, not only on the surface. It can be measured by RLCG bridge or DC power source (showing voltage and current values).

5.2. Resistance Modelling

The textile material can be modelled as finite grid of resistors. However, it assumes only woven fabric [11]. The example is depicted in Figure 10.

Measurement of resistance (bulk or surface) is based on placing two square electrodes on two ends of the woven textile material. The structure can be interpreted as series-parallel connection of resistors. The battery represents electrodes and resistors the textile fibres.

The equivalent circuit diagram can be simplified with respect to basic physical laws. Equipotential points in this diagram are in all individual „vertical" resistor connections. The resistors placed between the points with same potential can be eliminated because they are equalled to zero. The voltage probes are placed in the equipotential points in Figure 11. The results are depicted in Figure 12. All probes reach the same value and therefore the resistors can be eliminated.

The resultant resistance of this model can be calculated as series-parallel connection of resistors as:

$$R = \sum_1^{12} \frac{R1}{6} = \frac{12\,R1}{6} = 2R1 \tag{3}$$

Formula (3) can be generalized as:

$$R = \sum_{n=1}^{r} \frac{R1}{s},\ n, r, s \in \mathrm{N} \tag{4}$$

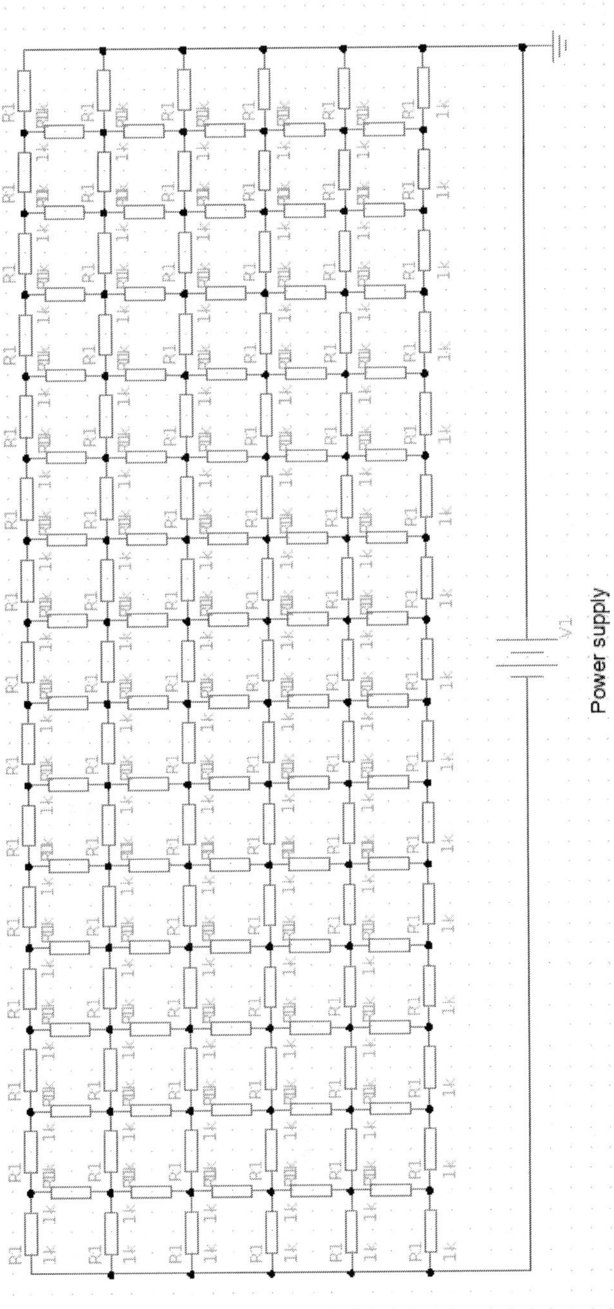

Figure 10. Equivalent circuit diagram of Betex textile material

where *n*, *r* represents number of squares in „horizontal" direction and *s* in „vertical" direction.

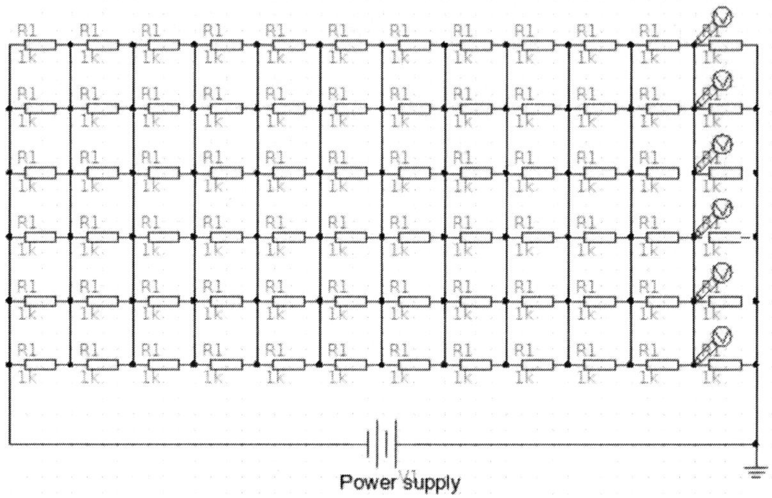

Figure 11. Simplified equivalent circuit model

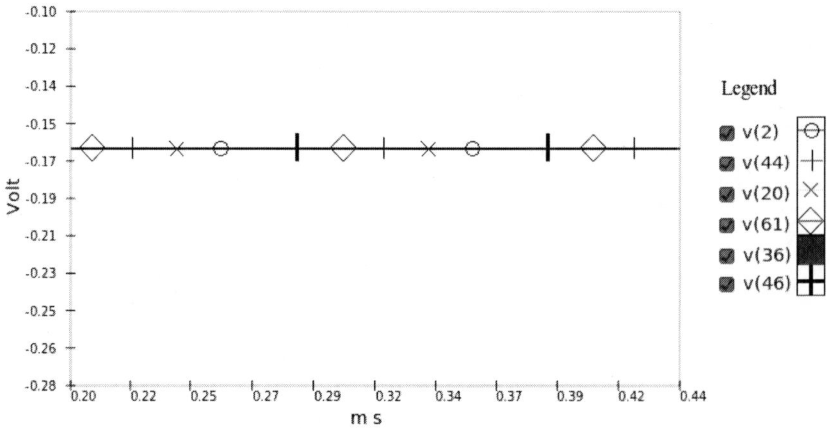

Figure 12. Result values of placed voltage probes.

Considering Betex sample and setup measurement, the Betex sample reach dimensions 10 x 3 cm, 25 threads/cm in warp and 20 threads/cm in weft. Parameters *n* and *s* are then equalled to:

$$r = 10 \cdot 25 - 1 = 249 \tag{5}$$

$$s = 3 \cdot 20 - 1 = 59 \tag{6}$$

Resultant resistance is equalled to:

$$R = \sum\nolimits_{n=1}^{r} \frac{R1}{s} = \sum\nolimits_{n=1}^{249} \frac{R1}{59} = \frac{249R1}{59} = 4.22R1 \tag{7}$$

The parameter *R1* represents a resistance element of used fiber which forms the whole fabric. It can be calculated from the dimensions of textile structure with the aid of fiber diameter measurement. *R1* is set to 0.97Ω and *R=4.09Ω.* It meands the structure is very conductive.

5.3. Resistance Measurement

The Betex sample is measured by RLCG bridge with respect to its calculated resistance. DC power source can cause sample damage at low voltage values (10 V corresponds to approx. 10 A). The measurement setup is depicted in Figure 13.

The measurement of Betex sample shows the resultant resistance is approx. 4Ω which confirms the modelling results.

6. SIMULATIONS

In this chapter different two antenna types are designed for the conductive textile material.The two designed antennas are H-slot antennas for RFID having an IC chip EM4222 at 869 MHz with the impedance 128-j577, microstrip patch antenna for RFID having T-match. Manual calculations and simulation results for all the antennas are presented below.

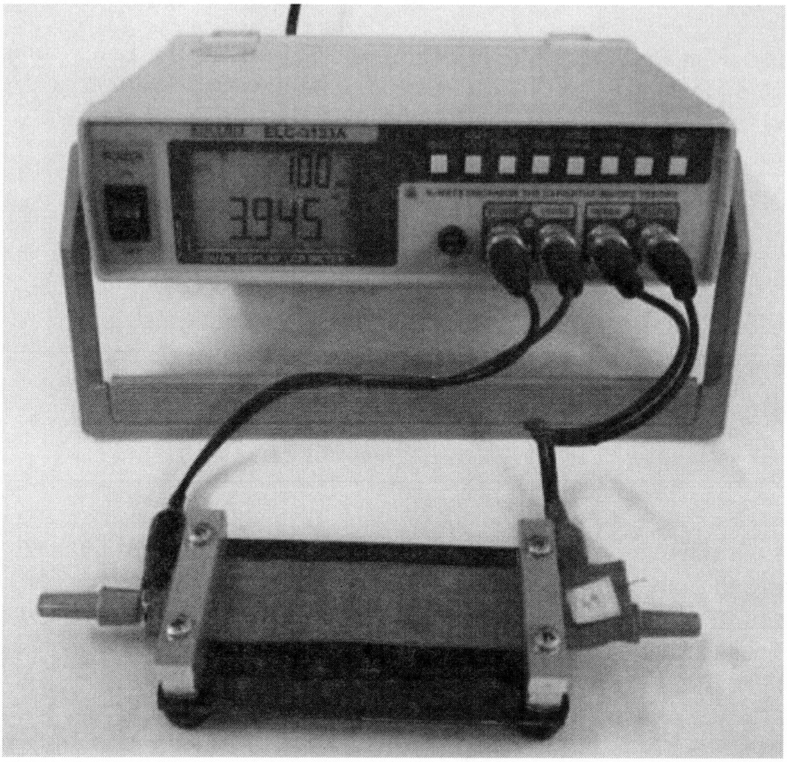

Figure 13. Measurement setup

6.1. Simulation Of H-Slot Patch Rfid Tag Antenna

The wearable tags are designed on IE3D, fabricated and tested in real conditions. The overall size of the H-slot antennas is 180 mm x 200 mm. This big dimension of the antennas can be smaller by using a substrate which has a high dielectric constant, because the antenna size depends on the dielectric constant ε_r of the substrate and also the design frequency. In this design a fleece fabric is used as a substrate which has a dielectric constant of 1.25. This fabric is chosen because of its better radiation performance. When a substrate has low dielectric constant and small thickness then the designed antenna has good radiation performance. But if a small tag is requested then a substrate with high dielectric constant can be used.

Three different substrate materials are used for comparison. The first substrate is Polyethylene (ε_r = 2.25, thickness h =1.7 mm). This design gives smaller dimensions (145 mm x 160 mm) then fleece fabric.

Later on this substrate replaces with the fleece fabric to increase the radiation performance because of fleece's low dielectric constant.

Third design is made by a substrate material which has a very high dielectric constant ε_r, to reduce the antenna size. The material is silicone slab. Silicone slab is chosen because it is elastic, hydrophobic material and this property gives an advantage of avoiding the water absorption into the substrate, this is important property of silicone slab because when water absorbed into the substrate, the dielectric constant of the substrate changes. This also gives homogeneous connection between the substrate and radiating patch. The antenna design with silicone slab is giving as a size of 57 mm x 78 mm, much smaller than other designs. But a decreasing in the radiation performance is achieved. Thus there is a trade of between antenna performance and size of the antenna.

In this work, two different conductive textiles are used to design and fabricate two different tags, TAG1 and TAG2.

Table 5. Specifications for the conductive textile radiating element

Parameter	Cu-Ni	BETEX
Frequency	869 MHz	869 MHz
Surface resistivity	0.02 Ω/sq.	1.19 Ω/sq.
Thickness	0.14 mm	0.35 mm
Conductivity	357143 S/m	2381 S/m

Table 6. Specification for simulating antenna (TAG1 and TAG2)

Parameter	TAG1	TAG2
Conductive Material	Cu-Ni, conductivy of 357143 S/m	BETEX, conductivity of 2381 S/m
Substrate Material	Fleece Fabric	Fleece Fabric
Thickness of the Substrate	4 mm	4 mm

6.1.1. Antenna Layouts and Designs

Cu-Ni conductive textile has high conductivity and the designing with this conductivity gives better radiation performance then Betex textile which has low conductivity. The simulation of two antennas with two different materials having same dielectric substrate and the same dielectric constant is shown in Figure 14.

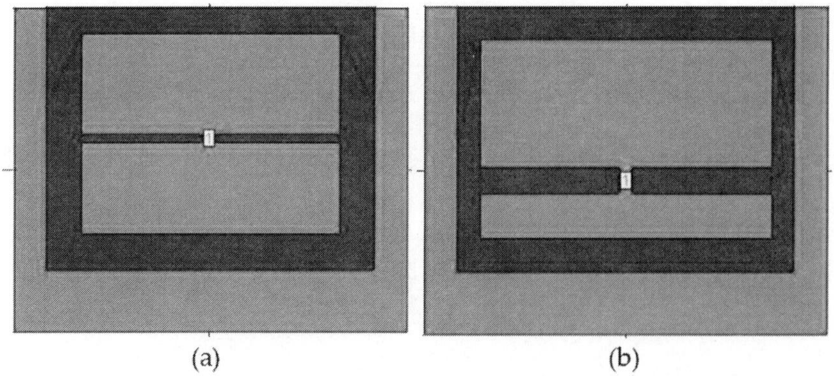

Figure 14. Simulated H-slot antenna for radiating element having surface resistivity = 0.02(a), and surface resistivity =1.19(b)

As expected, the tag with high conductive material is giving better radiation performance. When a comparison made between this two tag's simulations, a high reading distance and high radiation efficiency are achieved from TAG1 which is designed with high conductive textile. This can be shown by the plot obtained from the simulation, Figure 15.

It can be seen that radiation efficiency obtained for TAG1 is 57.1% and that for TAG 2 is 3.05%. As it is concluded before that the radiation efficiency is directly proportional to the surface resistivity of the radiating patch. Therefore, higher radiation efficiency is obtained for the TAG having lower surface resistivity. If the antenna were designed from copper material the simulated efficiency would be very high because copper has high conductivity then these conductive textiles. This is shown in later case. Therefore, this value of 57% radiation efficiency is quite good result for this conductive textile.

The value of CMF is used to find how good the matching is in between the antenna impedance and chip impedance. The plot obtained for TAG 1 and TAG 2 is depicted in Figure 18.

From the above Figure it can be observed that the CMF for TAG1 is 0.975 and that for TAG2 is 0.874. It can be considered that both of the tag antennas has good matching with the chip, however TAG 1 shows better match among the two.

6.1.2. Reading Range Calculation

The read range is obtained as:

$$d_{max} = \frac{c}{4\pi f}\sqrt{\frac{EIRP_R}{P_{chip}}\tau G_{tag}} \tag{8}$$

$$\tau = \frac{4R_{chip}R_a}{\left|Z_{chip}+Z_a\right|^2} \leq 1 \tag{9}$$

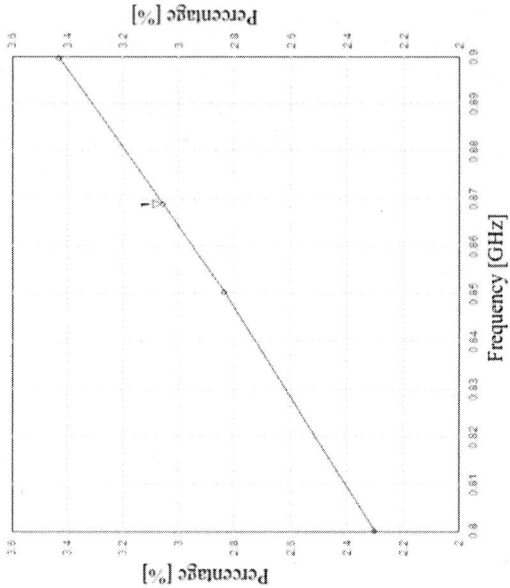

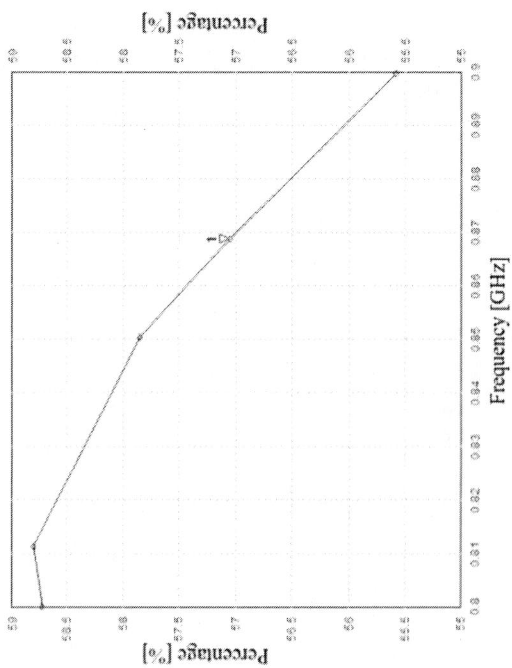

Figure 15. Radiation efficiency Vs Frequency plot for TAG 1 and TAG 2 respectively

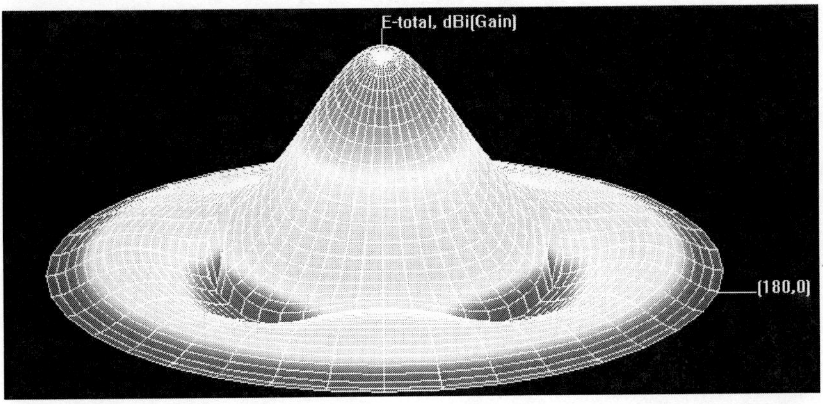

Figure 16. Pattern of TAG1

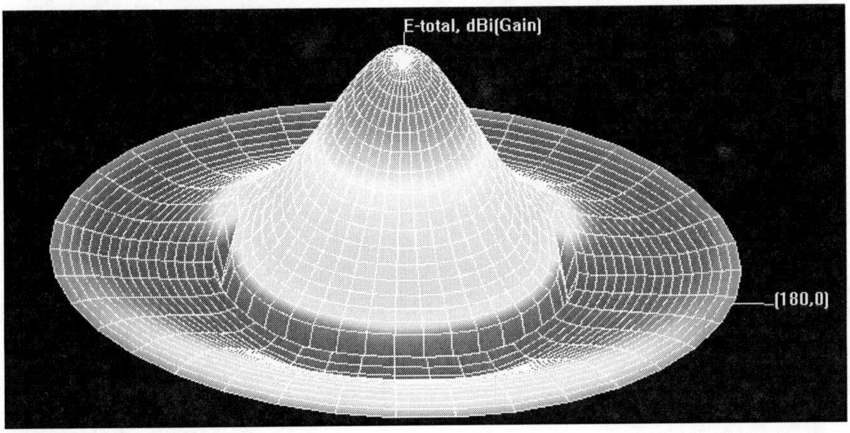

Figure 17. pattern of TAG2

Here, chip sensitivity is -10 dBm and the maximum radiated power by the reader is 3.2 W EIRP. Thus from the formula the transmission power coefficient for TAG1 and TAG2 is equal to,

$$\tau_1 = 0.97 \tag{10}$$

$$\tau_2 = 0.87 \tag{11}$$

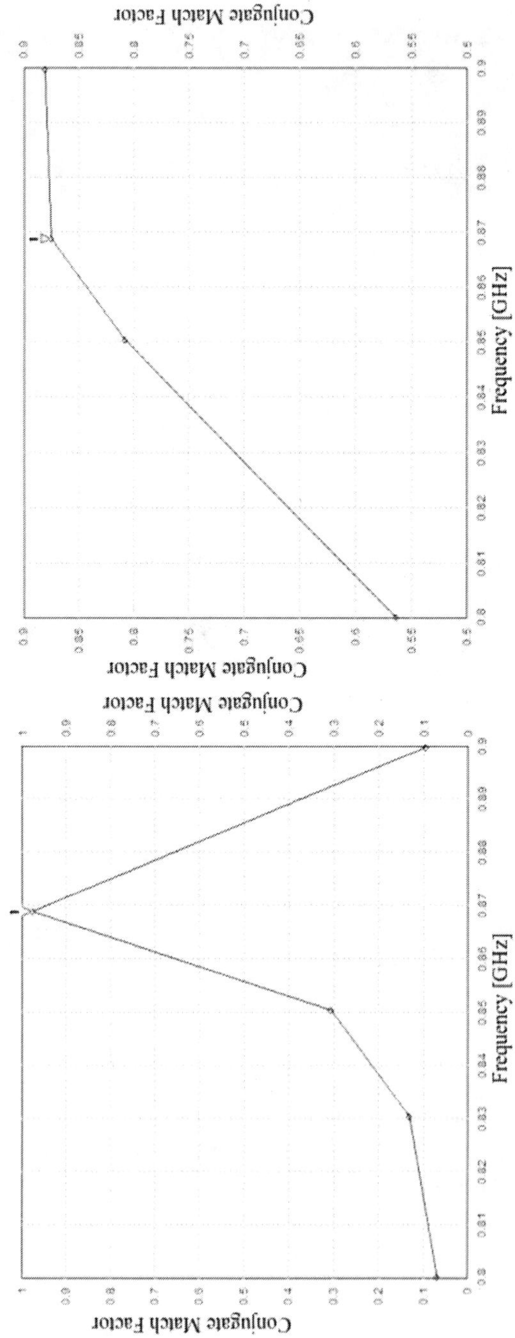

Figure 18. Conjugate match factor plot for TAG 1 and TAG 2 respectively

Thus the maximum range obtained is d_{max}= 2 m for TAG1 and d_{max}= 1.2 m for TAG2.

Table 7. The combined results obtained from the simulation of two tag antenna

@869Mhz	TAG1	TAG2
Radiation Efficiency,	57.1%	3.05%
Conjugate Match Efficiency	28.5%	1.52%
CMF	0.975	0.874
Gain	-8.46dBi	-15.25 dBi
Directivity	7.89dBi	8.47 dBi
Size of the Tag Lg x Wg	180x200mm	180x200mm
Antenna impedance, Za	100+j597	206+j472

6.2. Comparison When Different Dielectric Substrate Used

A comparison is performed in simulation to compare the radiation efficiency of H-slot antenna, when the conductive textile (Cu-Ni) is used as radiating element. The comparison is made by changing the thickness of dielectric substrate (fleece fabric) to 2 mm, 2.56 mm and 4 mm. The simulation is performed for the conductor having the surface resistivity

0.02 Ω/sq to resonate at the frequency 869 MHz. For three different thickness values, three different antenna geometry and CMF and radiation pattern is achieved. The CMF for all the designs were measured to be more than 0.95 when measured in linear scale.

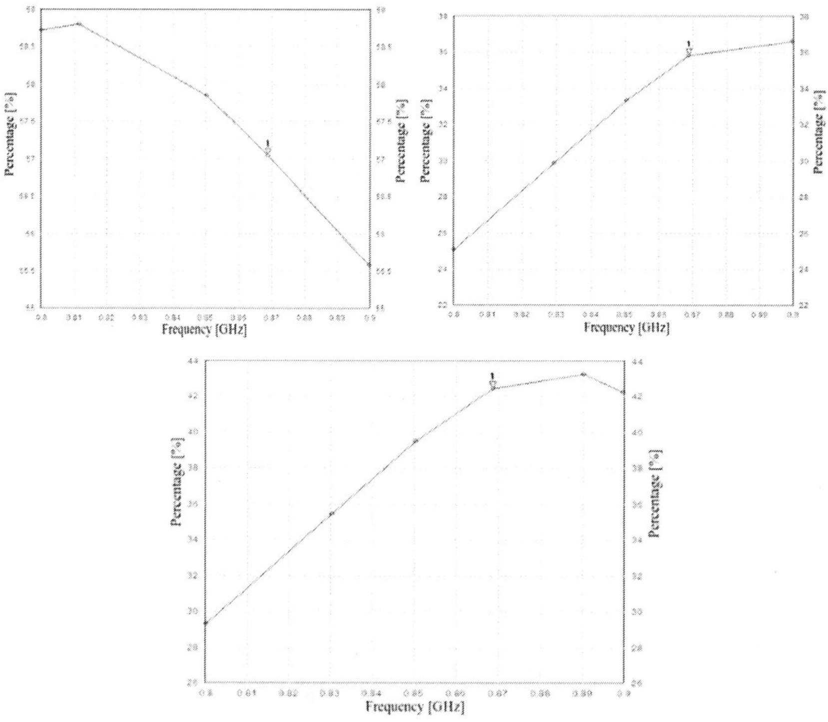

Figure 19. Simulation result of Radiation efficiency plot for Cu-Ni, with the height of dielectric substrate 2 mm, 2.56 mm and 4 mm.

Figure 19 depicts the radiation efficiency obtained from 4 mm thick dielectric substrate is the highest and obtained to be 57.1 %, for 2.56 mm thick dielectric substrate is 43.4 % and that for 2 mm thick dielectric substrate is 35.8 %. The better performance is achieved with the antenna having thicker dielectric substrate.

6.3. Comparison When Different Dielectric Material Used

In this case the two dielectric materials are used. One is the fleece fabric with the dielectric constant 1.25 and the other is silicone slab with dielectric constant 11.9. The silicone slab with high dielectric constant is used to reduce the size of antenna and also hydrophobic in nature. This is very useful characteristics of silicon slab. The two antennas with two different materials are depicted in Figure 20 and 21.

Figure 20. Antenna with Dielectric substrate fleece and antenna with dielectric substrate silicone slab respectively

As seen from the graph, the radiation efficiency of antenna using silicone slab and having higher dielectric constant, is reduced compared with the one using fleece fabric. Though the size of antenna was reduced, the efficiency was also decreased drastically. It is due to this reason fleece fabric is preferred.

6.4. Simulation Of Microstrip Patch Antenna For Rfid Application

This is another technique of designing RFID tag antenna. A rectangular parch antenna is used as a tag antenna for RFID. The microstrip patch antenna for RFID is designed for 869 MHz. The manual calculation of microstrip patch is calculated in the similar ways as for the rectangular microstrip patch in chapter 3, however the feeding is different. A T-match is used to match the impedance of the antenna to the chip. The calculated length is 150 mm and the width is 190 mm.

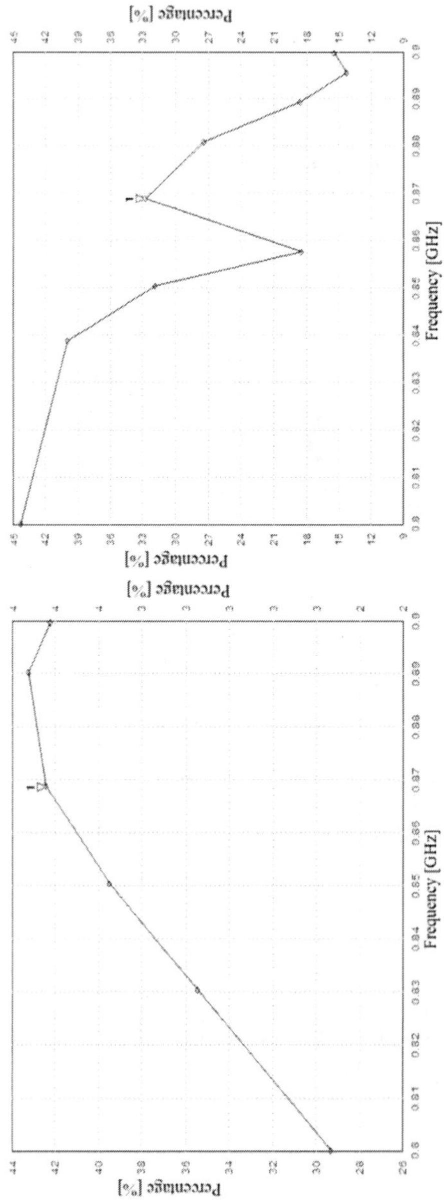

Figure 21. Radiation efficiency of antenna using dielectric substrate fleece (43.4%) and silicone slab (32.7%)

The dielectric material used is Fleece fabric with dielectric constant 1.25 and dielectric height of the substrate 2 mm. The radiating element is simulation of conductive textile material having surface resistivity 0.02 Ω/sq.

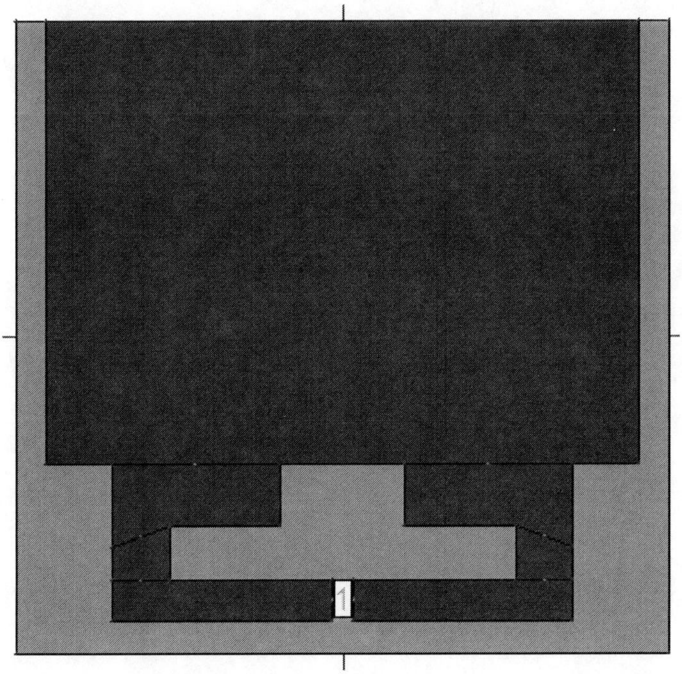

Figure 22. Microstrip patch antenna using T match

Table 8. Simulated output results for TAG 3

Frequency(869 MHz)	TAG3
Radiation Efficiency	41.82%
Conjugate Match Efficiency	20.91%

CMF	0.956
Gain	-7.99dBi
Directivity	8.162dBi
Size LxW	150x190
Imput Impedance , Za	92+j547
Reading distance	2m

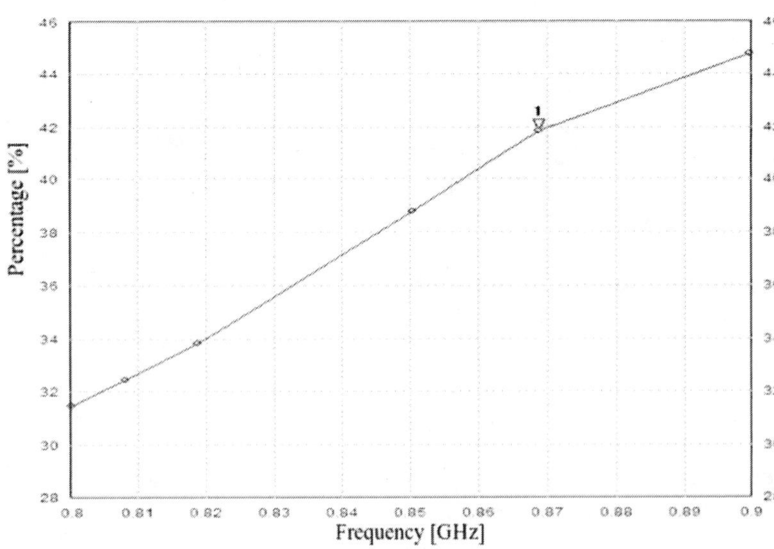

Figure 23. Radiation Efficiency Vs frequency plot for the tag antenna

Figure 23 shows the radiation efficiency obtained is 41.8 %.

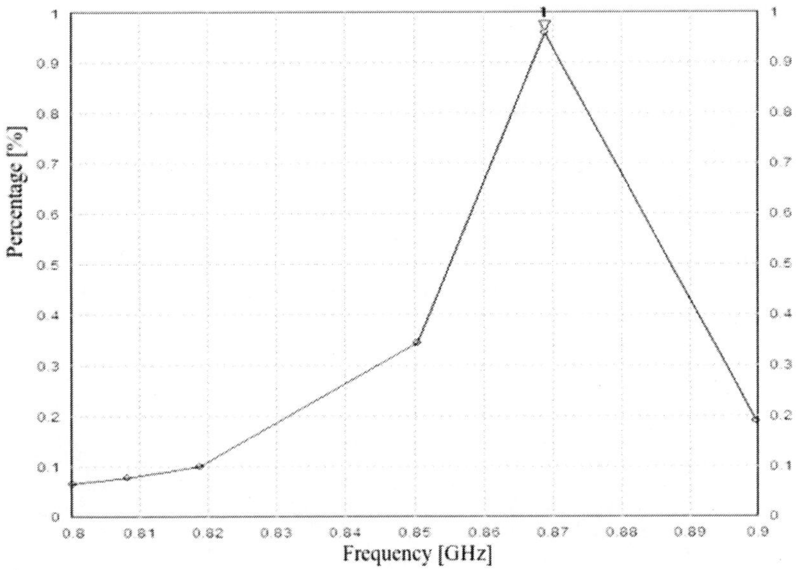

Figure 24. Conjugate match vs frequency for TAG 3 (CMF_{max}=0.956)

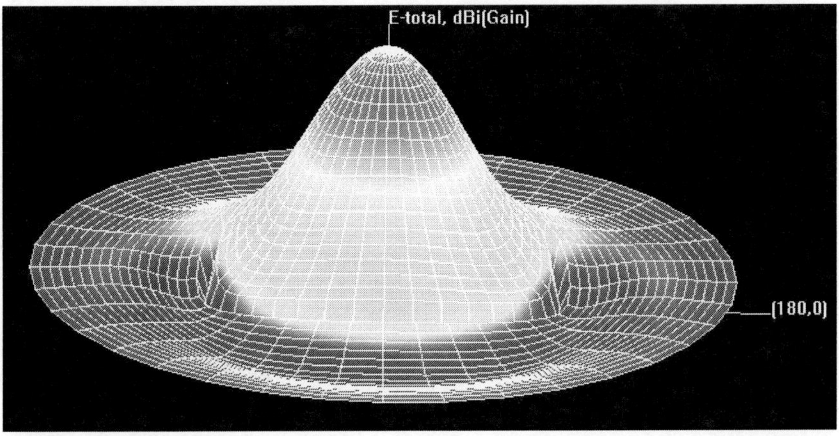

Figure 25. Radiation pattern for TAG 3

7. FABRICATION AND MEASUREMENT OF H-SLOT AND PATCH RFID ANTENNA

7.1. Results Of Fabrication and Measurements Of Tag1 And Tag2

For fabrication of these two tag antenna, the available fleece fabric had a thickness of 2 mm, so to have 4 mm thickness two layer of fleece is overlapped by using glue. The conductive material also attached to the substrate by using glue.

Figure 26. Fabricated antenna (a), chip connection to slot arm (b) for TAG 1

Figure 27. Fabricated antenna for TAG 2 and chip connection

As can be seen from the Figure 27 TAG 2 is constructed from Betex and being very difficult to connect chip by soldering, an alternative way is used. First a copper tape is attached, similar to that with microstrip patch and then the chip is soldered on top of it as shown in Figure. This not done with TAG 1, as it was not necessary.

Figure 28. Fabrication and Chip connection for TAG 3

To measure the tag performance, an RFID reader is connected to the computer.

The reader shown in the above Figure generates the frequency signal which is captured by the tag antenna, and retransmit signal back to reader. This signal is received by the reader and is sent to the computer for further processing of the signal.

The RFI21 RFID Reader Demo application program is used in the computer to read the reader. This application uses Python 2.6 programming language in the computer to accomplish this task. The reader is manufactured by METRA BLANSKO a.s.

During the measurement, the reading distance of the TAG1 and TAG2 is measured. Tags are moved towards the reader's antenna till the reader detects the signal from the tag.

As soon as RFID tag is detected by reader's antenna, the information is displayed on the computer.

Figure 29. Measuring Reading distance of TAG1

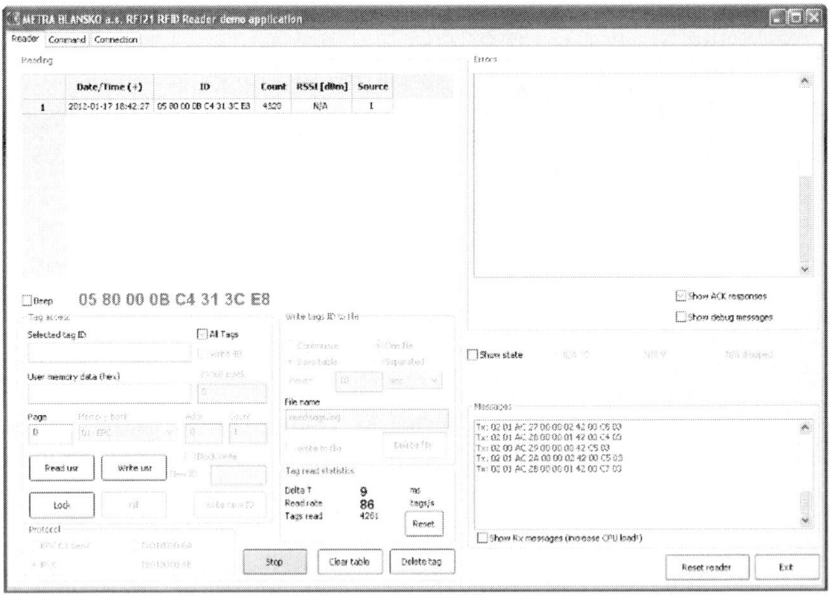

Figure 30. Application program detecting the EM4222 chip ID (in red)

EM4222 chip is used in the tag antenna. When the tag is detected by the reader the tag ID is displayed in the computer. This can be seen in the Figure which is the ID of the chip in red color.

From the measurement results, the reading distance for TAG1 is measured to be 50 cm. This is quite smaller then simulated results because of the change in dielectric material due to non-precise determination of permittivity and also soldering process.

The reading distance for TAG2 is quite close the simulation results, 90 cm. The simulated reading distance for this tag is 1.2 m. Thus TAG 2 gave better performance and was very close to simulated values.

Table 9. Measured read range from the three designed tag

Range	TAG1	TAG2	TAG3
Reading range	50 cm	90 cm	60 cm

A short experiment was done to compare working of the fabricated tag and the tag available in the market. To make a comparison of reading distance, two different UHF RFID tag's reading distance were measured. The tags used were UPM Hammer 258-1 and UMP short dipole 211_2. These are the commercially available tag in the market.

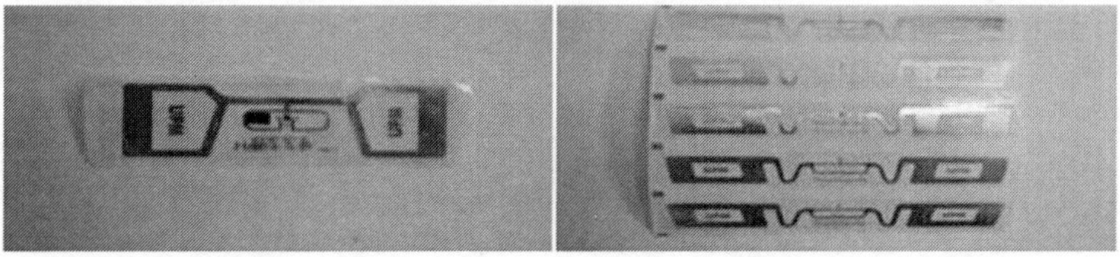

(a) (b)

Figure 31. UPM Hammer 258-1RFID tag (a), UMP short dipole 211_2 RFID tag (b)

8. CONCLUSION

Implementation of textile antennas for RFID tags represents a realistic developmental assignment, and as shown and practically proved, this arrangement yields good results. The successful operation of such textile antennas mainly requires the mechanical stability of the textile composite, which realizes a RFID tag antenna. A good function of the antenna and thus the sensitivity of the complete tag can be provided for just compliance with the mechanical construction and stability of required dimensions. A good choice of textile material for both electrically conductive structures and the insulating layer of the resultant fabric composite is the most important prerequisite for the successful implementation.

Table 10. Comparison of read range of manufactured tag with commercially available tag

Parameter	UPM Hammer 258-1 RFID tag	UMP short dipole 211_2 RFID tag	H-slot TAG2
Reading Distance	98 cm	152cm	90cm
Chip Protocol	EPS S1 Gento	EPS S1 Gento	IP-X

The measurement was performed in open space in the lab.

Textile RFID tags find its use at both person marking (marking of athletes, protective clothing and other functional ready-made textile products) and stock-in-trade marking in hospitals, packaging, etc.

ACKNOWLEDGEMENTS

This work was supported by the project Kompozitex FR– TI4/202 - Composite textile materials for humans and technology protection from the effects of electromagnetic and electrostatic fields. The work was conducted in the Department of Telecommunication Engineering at the Czech Technical University in Prague in the scope of thesis called "Design and performance analysis of purely textile antenna for wireless applications" [12] and other research projects.

REFERENCES

1. Kennedy T.F. et al. Body-Worn E-Textile Antennas: The Good, the Low-Mass, and the Conformal, IEEE Transaction on Antennas and Propagation 2009; 57(4) 910-918.

2. EM Microelectronic Data Manual, Marina SA. Company of the Swatch Group, Data Sheet EM4222, 2003.

3. Lozano-Nieto A. RFID Design Fundamental and Application. Boca Raton: CRC Press; 2011.

4. Occhiuzzi C. et al. Modeling, Design and Experimentation of Wearable RFID Sensor Tag, IEEE Transactions on Antennas and Propagation 2010; 58(8) 2490-2498.

5. Aniolczyk H. et al., Application of Electrically Conductive Textile as Electromagnetic Shields in Physiotherapy, Fıbers and Textiles in Eastern Europe 2004; 12(4) 47-50.

6. Vojtěch, L. and Neruda, M. Application of Shielding Textiles for Increasing Safety Airborne Systems - Limitation of GSM Interference. In The Ninth International Conference on Networks (ICN 2010). Los Alamitos: IEEE Computer Society 2010, 157-161.

7. Constantine A.B. Antenna Theory Analysis and Design. New York: John Wiley; 1997.

8. Maryniak W.A., Uehara T., Noras M. A. Surface Resistivity and Surface Resistance Measurements Using a Concentric Ring Probe Technique, Trek Application Note 2003; 1005: 1–4. www.trekinc.com/pdf/1005_Resistivity_Resistance.pdf (accessed 16 August 2012).

9. IEC 61340-5-1 Standard. Electrostatics – part 5-1: Protection of Electronic Devices from Electrostatic Phenomena - General Requirements, 2001.

10. ASTM Standard D 257-99. Standard Test Methods for D-C Resistance or Conductance of Insulating Materials, 1999.

11. Neruda, M. and Vojtěch, L. Verification of Surface Conductance Model of Textile Materials. Journal of Applied Research and Technology 2012; 10(4), 579-585.

12. Dahal, R. and D. Mercan. Design and performance analysis of purely textile antenna for wireless applications. Sweden, 2012. 64 p. Diploma thesis. Czech Technical Univesity, University of Gävle, Sweden. hig.diva-portal.org/smash/get/diva2:504525/ FULL TEXT 01 (accessed 16 August 2012).

6

INTEGRATING RFID WITH IP HOST IDENTITIES

Steffen Elmstrøm Holst Jensen[1] and Rune Hylsberg Jacobsen[1]

[1] Aarhus University School of Engineering, Denmark

1. INTRODUCTION

The "Internet of Things" semantically means "a world-wide network of interconnected objects uniquely addressable, based on standard communication protocols" [1]. The vision describes a world that enables physical objects to act as nodes in a networked physical world [2]. The terms "Internet of Things" can be attributed to The Auto-ID Labs, a world-wide network of academic research laboratories in the field of networked RFID and emerging sensing technologies [2]. Together with EPCglobal®, these institutions have been architecting the Internet of Thing since their establishment. Their focus has primarily been on the development of the Electronic Product Code™ (EPC) to support the wide-spread use of RFID in modern, global trading networks, and to create an industry-driven set of global standards for the EPCglobal Network.

EPCglobal Network was created for "traditional" low-cost tags [3]. The main functionality of the EPCglobal Network is to provide data assigned to a specific tag, so that each RFID read event can be stored in a database and applications can be built on this data. Since tags were not originally considered to carry or compute additional data, the EPCglobal Network does not traditionally provide a mechanism to address remote tags from networked applications.

The data flow in these networks works from tags via readers to a couple of networked servers. Passive, low-cost RFID tags are widely available and the EPCglobal Network was defined to support open-loop supply chain applications. Basically, this is accomplished by allowing servers to communicate over the Internet. Although RFID technology is quite accepted in closed-loop applications, the evolution towards open-loop systems using the EPCglobal Network with distributed databases did not take place as predicted due to problems in the access control layer of such systems.

RFID-sensor networks are an emerging part of the Internet of Things [5]. These devices combine sensing capabilities with an RFID interface that allow the retrieval of sensed data. In fact, they can cooperate with RFID systems to better track the status of things e.g., their location, temperature, movements, etc. A sensor-enabled RFID tag (also known as sensor-tags) is an RFID tag which contains one or more sensors to monitor some physical parameter (e.g., temperature) but also contains the same identification function as a "normal" RFID tag does. This kind of sensor tag may fall into class 2, class 3 or class 4 in EPCglobal's tag classification [3]. A fully passive, class 2 sensor-tag can measure physical parameters, i.e., use sensors, only when powered by a reader. In contrast, class 3 tags are battery assisted. They can work independently of the reader and can be suitable for RFID-sensor networks.

In this chapter, we will discuss different ways to achieve the Internet of Things vision by internetworking passive RFID tags over IPv6. The chapter is organized as follows: Section 2 presents related works and discusses the novelty of the work presented here. Section 3 introduces the key technologies for the convergence of RFID and Internet namespaces and to provide an address mapping needed to internetwork passive RFID tags. In Section 4, some common examples of RFID usage are given and discussed in the context of globally networked tags. Subsequently, Section 5 introduces a testbed built to study the interconnection of passive RFID tags over IPv6. The different strategies that can be used for integrating RFID with IPv6 are discussed in Section 6 and this discussion is followed by mobility considerations in Section 7. Finally, Section 8 concludes the discussion and outlines anticipated future work in this area.

2. RELATED WORKS

Most objects in our surrounding are not equipped with microprocessors and hence cannot attach to a computer network. However, these objects can be equipped with passive, low-cost RFID tags either as tags integrated or adhesively stuck to the object and hereby provide a mean of communications. Dominikus et al. [14] has suggested a way to integrate passive RFID systems into the Internet of Things, by using readers that function as IPv6 routers. In their work, an IPv6 addressing scheme that map tag IDs to network addresses was defined. Furthermore, the mobility problem, which arises when tags physically moves around, was investigated and the use of Mobile IPv6 (MIPv6) to cope with tag mobility was suggested. In contrast to the work presented by Dominikus et al. [14], this chapter opens the discussion on the proper formatting of the IPv6 addressing by introducing cryptographic hashing techniques as well as the possibility of separating identity and location information when forming an IPv6 address. The use of hashing techniques to construct an IPv6 address from an EPC, as opposed by using a compressed EPC format [14], eases practical implementations and allows the use of the same mapping scheme for all EPC types.

An alternative approach is to provide the tags themselves with the IPv6 protocol stack, making them able to use IPv6 communication over the Internet whenever close to a reader. This requires several changes to the design of existing tags. In this case, the tags do all the work themselves and need a separate power source. A solution where the tags are modified to hold the IPv6 stack on them is discussed by Rahman et al. [4]. The tags EPC, which is its identity, would then be made into a part of the tags IPv6 address due to the design of the tags proposed. This makes these tags too expensive for integration into the Internet of Things since the price of the tags could easily exceed the value of the "things" themselves.

Barish et al. [13], describes a somewhat similar setup than the one proposed here. In their approach, a global address manager is used to keep track of tags. The basic idea is that an application sends the EPC to a global server along with the IP address that the tag has been associated with. When a corresponding node wants to communicate with the tagged object, it contacts the last known address. If the tag is in the field of the reader the connection is established and communication can begin. If the tag is not present at the location the request is redirected to the global address server that returns the tag's present address or just redirects the request to the correct address. In contrast to the proposed solution by Barish et al. [13], the approach described here does not include extra nodes in the network to construct network addresses but adds functionality to the RFID readers residing at the network edge.

Xu et al. [25] proposed a general address mapping scheme based on a proprietary protocol named General Identity Protocol (GIP). The scheme takes all existing RFID systems into account, and allows heterogeneous RFID systems to interwork over the Internet. This is accomplished by mapping RFID tag identifiers to IPv6 addresses, constructing a GIP message with details of the RFID systems in use, and finally encapsulating the message in IPv6 and routing the packet over the Internet. This chapter describes a solution that minimizes the need for control protocols.

3. ENABLING TECHNOLOGIES

There are a couple of ways to interconnect objects by using RFID with IPv6 [6]. One solution would be to give the tags the ability to communicate via the Internet. The communication can be both reader-initiated and tag-initiated. The latter requires specific tags that require electrical and processing power to be available in the tag such as e.g., EPC class 3 tags. Most of the computational work takes place in the tags, i.e., the tag is reachable and visible as an IPv6 connected host as long as it is within the electric field of a reader.

Passive RFID tags, such as EPC class 2 tags, do not have the possibility to power a network protocol stack and therefore a network address cannot be directly assigned to the tag's microchip. However, the passive tag can be represented by virtual interfaces residing in the reader interrogating the tag.

3.1. Radio Frequency Identification (Rfid)

RFID systems are composed of one or more readers and several electronic tags. Tags are characterized by a unique identifier that takes the form as a binary number. They are applied to objects and even persons or animals as implants. From a physical point of view, an RFID tag is a small microchip attached to an antenna that is used for both receiving the reader signal and transmitting the tag ID. The dimensions of each tag can be very small with tag dimensions down to 0.05 mm x 0.05 mm with a thickness of 0.005 mm [7]. There are more than 60 tag manufacturers world-wide [8].

RFID tags will act as electronic identification for physical objects to which they are linked. In the Internet of Things, all objects, virtual as well as physical, are interconnected and reachable via for example IPv6 in combination with RFID technology [6]. Essentially, the tag connects to physical objects that we want to authenticate and track when they come in contact with readers. A reader can read or modify tag's information. The back-end database keeps information related to different tags/readers.

For reader-initiated communication the reader triggers the tag's transmission by generating an appropriate signal, which represents a query for the possible presence of tags in the surrounding area and for the reception of their identification codes (IDs).

Active tags come with a power source that can drive a microprocessor (or microcontroller). Furthermore, it allows a stronger electromagnetic field to be generated in response to an incoming RFID air protocol message and larger read distances can be achieved. More advanced active tags or sensor-tags may run additional software and can be equipped with communication software such as an IP protocol stack [9].

In contrast, passive tags rely on the incoming electromagnetic field from the reader to power the circuit and to deliver power to drive the response to an interrogating request. These devices do not run communication software and cannot be actively involved in a protocol message exchange. To communicate with these devices there is a need for software agents to act on their behalf.

RFID tags can only be "online" when they are in the electric field of a reader field. For high velocity applications, where tags only remain certain seconds in a reader field, the proposed approach of networking these tags is not applicable.

3.2. Rfid Namespaces

Essentially, RFID comes with two namespace to be used with RFID applications: the EPC addresses and the Object Name Service (ONS). A namespace can represent objects as well as concepts and may be generalized as a container for a set of identifiers (names). The EPC is an identifier based on the standards established by EPCglobal® [10]. It is designed to allow the automatic identification of objects anywhere. EPC defines three layers of identity: the *pure* identity, the encoding layer identity and the physical realization of an encoding. The EPC tag data standard [10] identify how existing coding systems such as the GS1 family codes for serialized human readable representations e.g. GTIN, GCN, SSCC, GRAI, GIAI, GSRM, GDTI and a small number of other identities should be embedded within the EPC.

A canonical representation of an EPC is the *pure identity* Uniform Resource Indicator (URI) representation, which is intended for communicating and storing EPC in information systems, databases and applications. The purpose is to insulate EPCs from knowledge about the physical nature of the tag, so that although 64-bit tags may differ from 96-bit tags in the choice of binary header values and in the number of bits allocated to each element or field within the EPC, the pure identity URI format does not require the information system to know about these details. Hence, the pure identity URI can be regarded as a pure identifier [10]. Tags are identified by URIs such as e.g., urn: epc: id:sgtin:0523141.000024.120 that comprise both tag number and associated coding scheme.

Encoding is the process of translating the pure identity EPC into a specific instantiation incorporated into tags for a specific purpose. During the encoding process the URI information is translated into a binary encoding that is stored in the tag. Subsequently, translating between the different levels of representation can be accomplished in a consistent way.

Figure 1 shows the structure of the data layout of an EPC code for the Global Trade Item Number (GTIN) and Serialized Global Trade Item Number 96-bit (SGTIN-96) tag.

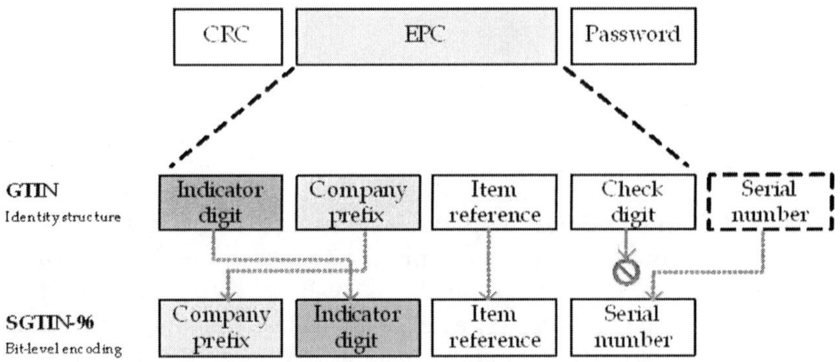

Figure 1. Tag data layout example.

EPC generation 1 standards, i.e., class 0 and class 1 tags, use a Cyclic Redundancy Check (CRC) to verify data integrity and a password as a "kill code" to disable the tags. The password must never be transmitted under any circumstances [3]. The *Item reference* identifies a class of objects to be tagged and it allows the grouping of items. The *Serial number* identifies an instance of a particular item. Company prefixes (also known as General Manager Numbers) point to the organization responsible for the subsequent partition. Finally, *Indicator digits* are used to specify length, type, structure, version, and generation of the EPC. This latter part is further used to guarantee uniqueness in the EPC namespace. For the GTIN encoding a *Check digit* is used.

Since the EPC is the only required data stored on a tag, it must be used as a "pointer" to find additional data about an object to which it attaches. This additional data should be stored on a server connected to the enterprise network or to the Internet. The server is identified via a look-up system which is called ONS.

ONS acts as a directory service for organizations wishing to look up product numbers (also known as EPC numbers) on the Internet. The ONS is operated as part of the EPCglobal Network. It is based on the well-known DNS service and it realizes the link between EPC numbers and EPC Information Services (EPCIS) as illustrated in Figure 2. When an RFID reader reads a tag, the EPC is passed to a middleware which then looks the EPC up either on the local machine, or enquires ONS through the Internet.

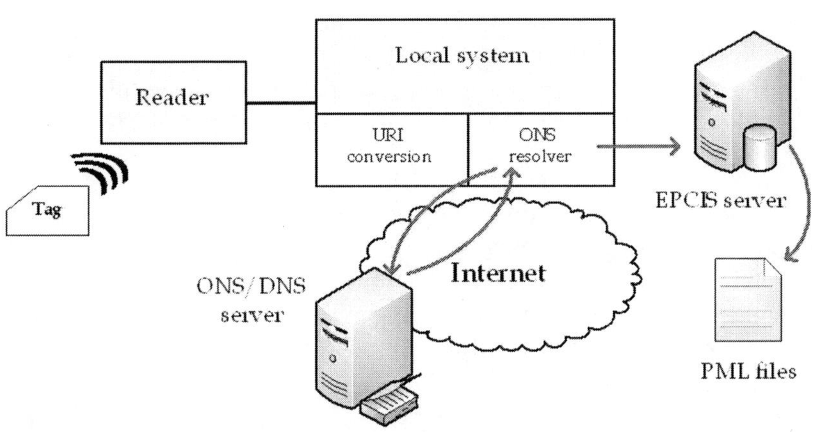

Figure 2. The Object Name System (ONS). Adapted from [**11**].

The ONS resolution process takes the EPC code and returns network location(s) where information resides, i.e., the EPCIS server, which typically holds web pages with information about tags. The Physical Markup Language (PML), based on XML technology, is intended to be the standard in which information about tags should be written.

In contrast, the DNS of the Internet will handle many more requests in the future. Therefore, enterprises will likely maintain ONS servers locally, which will store information for quick retrieval. Hence, a manufacturer may store ONS data from its current suppliers on its own network, rather than pulling the information off a Web site every time a shipment arrives at the assembly plant.

3.3. Internet Namespaces

There are two principal namespaces in use in the Internet: IP addresses and domain names. Domain names provide hierarchically assigned names for some computing platforms and some services. Each level in the hierarchy is delegated from the level above. Email, Hypertext Transfer Protocol (HTTP), and Session Initiation Protocol (SIP) addresses all reference domain names to mention its most wide-spread use.

On the network layer, IP addresses are used. IPv6 was introduced in the 90'ies due to the foreseen lack of globally unique IPv4 addresses, resulting in a protocol specification released in 1998 [17]. The IPv6 address is a 128-bit address that takes the form of a 64-bit network prefix appended by a 64-bit host suffix/interface identifier. The network prefix is used for routing purpose and determines the location of the host in the Internet. The host itself is identified by an interface ID. Figure 3 shows the IPv6 address format and gives an example on how a 96-bit EPC can be mapped to a network address.

It can be observed form the figure that not all 96 bits of the EPC can be fitted within the host suffix/interface identifier of an IPv6 address. Because of this deficiency, the implementers of an RFID-to-IPv6 mapping scheme is faced with a number of design options. These options basically govern the strategy for the mapping and are the subject of our discussion in Section 6.

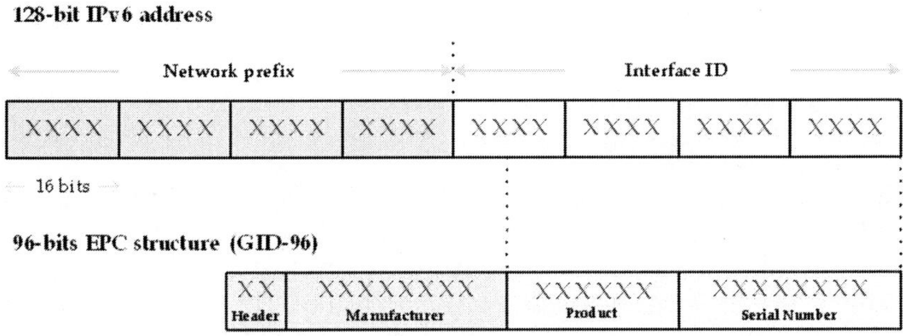

Figure 3. IPv6 address format compared to a 96-bit General Identifier (GID-96) EPC format. An 'X' indicates a grouping of 4 bits.

3.4. Cryptographically Generated Addresses

Cryptographically Generated Addresses (CGAs) are IPv6 addresses for which the host suffix/interface identifier is generated by computing a cryptographic one-way hash function from a binary input such as e.g., a public key [27]. CGAs are intended to be globally unique in a statistical sense but these may not necessarily be routable addresses at the IP layer [12].

The Overlay Routable Cryptographic Hash Identifiers (ORCHID) is a new, experimental class of identifiers based on CGAs. ORCHIDs have an IPv6-like address format and can be used with existing applications built on IPv6 [12]. These identifiers are intended to be used as pure endpoint identifiers for applications and Application Programming Interfaces (APIs) and not as identifiers for network location. This is in contrast to the IPv6 address that uses the 64-bit network prefix as locator [17].

While ORCHIDs use public cryptographic keys as input bit strings, it is possible to use the binary EPC encoding instead. The algorithm to generate an ORCHID in an RFID context is outlined below [12]. The algorithm takes a bitstring and some *context identifier* as inputs and produces an ORCHID output that is formatted as an IPv6 address.

$$Input := anybitstring \tag{1}$$

$$HashInput := ContextID \mid Input \tag{2}$$

$$Hash := Hash\_function(HashInput) \tag{3}$$

$$ORCHID := Prefix \mid Encode_n(Hash) \tag{4}$$

Concatenation of bitstrings is denoted '|'. The *Input* is a bitstring that is unique within a given context. The *Context ID* is a randomly generated value defining the expected usage context for the particular ORCHID and the hash function to be used for generation of ORCHID in this context. The purpose of a context ID is to be able to differentiate between various experiments that share the ORCHID namespace. The *Hash_function* is a one-way hash function to be used to generate ORCHIDs such as SHA1 [23] or MD5 [24]. SHA1 and MD5 produce a 160-bit and a 128-bit output, respectively.*Encoden* is a function to extract an *n*-bit-long bitstring from its argument. Finally, *Prefix* is an IPv6 network prefix.

To construct a CGA an input bitstring and context identifier are concatenated to form an input datum, which is then fed to the cryptographic hash function. The result of the hash function is processed by an encoding function, resulting in an *n*-bit-long output. This value is prepended with the network prefix resulting in a 128-bit-long bitstring identifier that can be used for programming with the IPv6 API.

To create a CGA namespace for RFID tags the EPC of a tag and the network prefix assigned to the reader that interrogates the tag are used as input. Furthermore, an $Encode_{64}$ function is used to extract 64 bits from the hash. A key advantage of using hash values over the actual raw host identity resulting from the EPC is its fixed length. This makes protocol implementations easier and it alleviates the management of packet sizes. However, a claimed drawback is that CGAs work one-way, meaning that it is not possible directly to create the original identity from the hash.

A CGA can be globally unique or globally unique in a statistical sense. That is, the probability of the same CGA being used to refer to different entities in the Internet must be sufficiently low so that it can be ignored for all practical purposes. Even though CGA collisions are expected to be extremely rare, collisions may still happen since it is possible that two different input bitstrings within the same context may map to the same CGA. A second type of collision can happen if two input bitstrings, used in different contexts, map to the same CGA. In this case, the main confusion is about which context to use. In order to preserve a low enough probability of collisions, it is required that applications ensure that distinct input bitstrings are either unique or statistically unique within a given context. By adhering to the EPCglobal standards, this requirement is fulfilled.

3.5. Host Identities and Host Identity Protocol

A host identity is an abstract concept assigned to a computing identity platform. In this section, we will generalize this concept to cover thin compact platforms that can be equipped with RFID tags. The discussion starts by introducing the host identities and the host identity protocol [15][16].

The Host Identity Protocol (HIP) supports an architecture that decouples the transport layer (TCP, UDP, etc.) from the internetworking layer (IPv4 and IPv6) by using public/private key pairs, instead of IP addresses, as host identities [15][16]. The public keys are typically, but not necessarily, self-generated. HIP introduces a new Host Identity (HI) namespace, based on these public keys, from which end-point identifiers are taken. Host identifiers are used to bind to higher layer protocols instead of binding to IP addresses. A key benefit of this approach is that it is compatible with existing APIs such as the socket API. HIP uses existing IP addressing and forwarding for locators and packet delivery, respectively.

Figure 4 illustrates the difference between binding of the logical entities service and end-points to an IP address (left side of figure). The service typically binds to the IP stack via the socket API. By using the host identity abstraction of the HIP architecture, the service and the end-point bind to the host identity whereas the location is still anchored with the IP address.

There are two main representations of the host identity, the full Host Identifier (HI) and the Host Identity Tag (HIT). The HI is a public key and directly represents the identity. The HIT is the operational representation of a host. It has a 128-bit long representation and is used in the HIP payloads to index the corresponding state of the end hosts. By introducing an identity concept at the network layer, where every host is represented by an asymmetric key pair consisting of a public and private key, it turns IP addresses into pure locators.

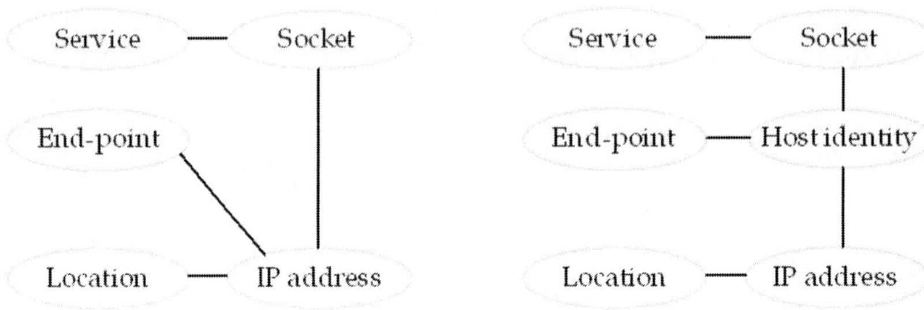

Figure 4. Illustration of the difference between the bindings of the logical entities. Adapted from [**15**].

The proposed HI namespace fills an important gap between the IP and DNS namespaces. A public key is used as the HIP Host Identity (HI), while the private key serves as proof of ownership of the public key. To seamlessly integrate HIP with protocols above the network layer, a 128-bit cryptographic hash of the HI, i.e., the HIT, was introduced to fit the IPv6 address space. The HIT is a statistically unique flat identifier. When HIP is used, the transport layer binds to HITs. In this process it becomes unaware of the IP addresses that are used for routing.

To be able to setup communication between peers that use HI, a light-weighted protocol exchange called the HIP Base Exchange has been specified. In Figure 5, the HIP Base Exchange is adapted to an RFID setup. The setup is somewhat similar to the one presented by Urien et al. [11].

Since the deployed tags are passive, there is a need for a proxy to act on behalf of the tags in the protocol exchange. The role of this proxy will be explained further in Section 5.

The HIP Base Exchange is a four-way handshake between two hosts wanting to initiate communication (see Figure 4). The *Initiate 1* packet is the first packet sent in the handshake. It is an unencrypted and unsigned packet, meaning that the Initiator would like to talk HIP with the Responder. The HIP packet contains the HIT of the Initiator (HIT_I) and the Responder (HIT_R). The responder's IP address can be derived from the DNS. *Respond 1* is sent as a reply to the *Initiate 1* packet. Besides the HIT_I-HIT_Ridentity pair, it contains a cryptographic puzzle challenge, and Diffie-Hellman parameters (DH_R) for the Diffie-Hellman key agreement. The Diffie–Hellman key exchange method allows two parties that have no prior knowledge of each other to jointly establish a shared secret key over an insecure communication channel.

Subsequently, the secret key can be used for encryption and integrity protection of the communication channel. The purpose of the HIP puzzle mechanism is to protect the Responder from denial-of-service attacks. The *Initiate 2* packet returns the corresponding Diffie-Hellman parameter (DH_I) to the Responder. It carries an encoded solution to the puzzle. Upon reception of the *Initiate 2* packet, the responder can now generate the keying material, and it is capable of using it in encryption and integrity protection algorithms. The Response 2 packet completes the HIP Base Exchange. After the Base Exchange, there is no longer difference between the Initiator and Responder and data can securely be exchanged between the communicating peers.

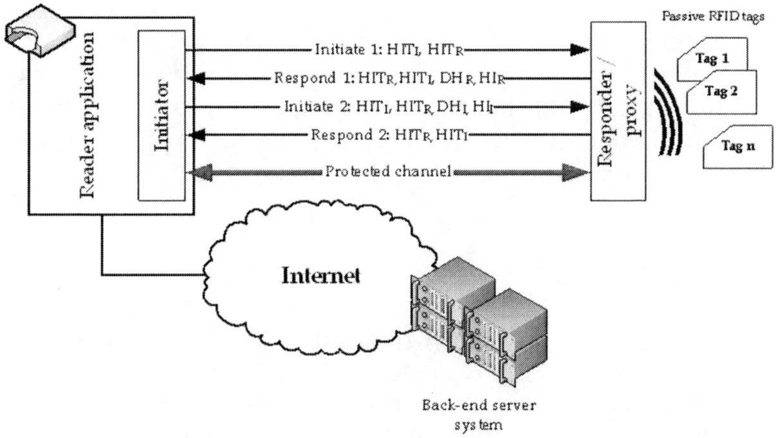

Figure 5. HIP Base Exchange adapted to a RFID communication scenario. Adapted from [**15**].

4. USE CASES AND APPLICATION EXAMPLES

RFID applications are numerous and far reaching [8]. The most interesting and widely used applications include those for security and access control, supply chain management, and the tracking of important objects and personnel. This section outlines a number of commonly encountered use cases for RFID technology, and discusses these in the context of networked RFID tags.

4.1. Access Control

Access control systems are an important part of the security of government buildings, companies, schools, residences and private areas and RFID technology has been widely adopted in access control systems. These systems often use RFID identification cards based on the IEC/ISO 14443 [18], IEC/ISO 15693 [19], or IEC/ISO 18000 standards [20]. The identification cards work much like a traditional key for unlocking doors or otherwise granting access. However, RFID technology does not provide authentication to the holder of the RFID card (or tag). Any unauthorized people holding an authorized RFID card could get access to secured area. Therefore, RFID technology should be combined with other means of identification such as e.g., face recognition to strengthen the security of the access control system.

By associating a passive RFID tag such as a key card with a globally unique IPv6 address we will be able to use access control and security policy mechanisms with Internet technologies to provide the desired access control applications. In this scenario a door locking mechanism would be connected over the Internet resulting in a more open system architecture.

4.2. Supply Chain Management

Most supply chain applications involve the concept of inventory tracking. An example of a proposed use of RFID is to ensure safety in the supply chain [21].

To illustrate the potential of using network RFID tags with supply chain applications an example taken from the Tag Data Standard v1.6 issue 2 [10]. The example text is quoted below:

"... a shipment arriving on a pallet may consist of a number of cases tagged with SGTIN identifiers and a returnable pallet identified by a GRAI identifier but also carrying an SSCC identifier to identify the shipment as a whole. If a portal reader at a dock door simply returns a number of binary EPCs, it is helpful to have translation software which can automatically detect which binary values correspond to which coding scheme, rather than requiring that the coding scheme and inbound representation are specified in addition to the input value."

Each of the cases tagged will be given a unique IPv6 address when they enter the electric field of a reader. This process involves the extracting of the essential bitstring of the SGTIN identifier for each case. Likewise, the returnable pallet and the shipment as a whole will be given IPv6 addresses that can be built based on the GRAI and the SSCC, respectively. By using the assigned IPv6 unicast addresses it is possible to establish communication to individual cases as well as the pallet. However, it may be of less interest to address individual cases at this point in the supply chain but rather to address the ensemble of cases. By introducing multicasting at the network layer it can be possible to communicate with groups of cases on the pallet.

4.3. Object/Asset Tracking

Because moving objects can easily carry RFID tags, a common use is to track the movement of people and the information associated with them. By associating a particular tag's EPC with a global network address the task of tracking the object/asset become equivalent to locating a mobile host in the network. In general, this is a key challenge in mobility research and several solutions have been proposed [22][26], and this will be the subject of our discussion in Section 7. Another interesting use case can be applied to sensor-tags. When these sensor-tags connect to a network sensor data can be retrieved from the tag.

5. NETWORKED RFID TESTBED

To study the internetworking of objects with passive RFID tags, a simple testbed has been built. The approach makes use of an RFID reader and an application that works as a proxy for the tags we wish to communicate with. The proxy is capable of making a virtual representation of the passive RFID tag on the Internet by creating a Virtual Network Interface (VNI) with an IPv6 address that can be attributed to each tag that comes within the electric field of a reader. Hence, the tags do not terminate IPv6 traffic directly but merely communicate with an entity which represents the tag (physical object) that we wish to communicate with.

The approach taken is software-oriented. The application runs as standalone but it can be embedded on the reader or it can be run on a computer local to the reader. The application receives the EPC of a tag attached to a physical object via the reader. The application then creates an IPv6 address from one of the mentioned methods. Hereafter, it is possible to route IP traffic to the particular Internet end-point. This will in effect make the application act as a proxy that for example can keep the most recently read tags "online". The solution gives a one-to-one mapping of physical objects to the virtual representations that are needed to communicate over the Internet.

Figure 6 illustrates the system implemented. In practice, the application host has a predetermined number of Virtual Network Interfaces (VNIs) installed. These interfaces work as the online virtual representation of the tag swiped at the reader. In other words, this is the interface the outside world can contact. In the testbed, the network interfaces are virtualized in a way similar to a loopback interface [17]. As the system works as a testbed the database is merely there as a logging service. In the future, it is planned to use the database as foundation for a local ONS. The corresponding node is there to illustrate possible communication over the Internet.

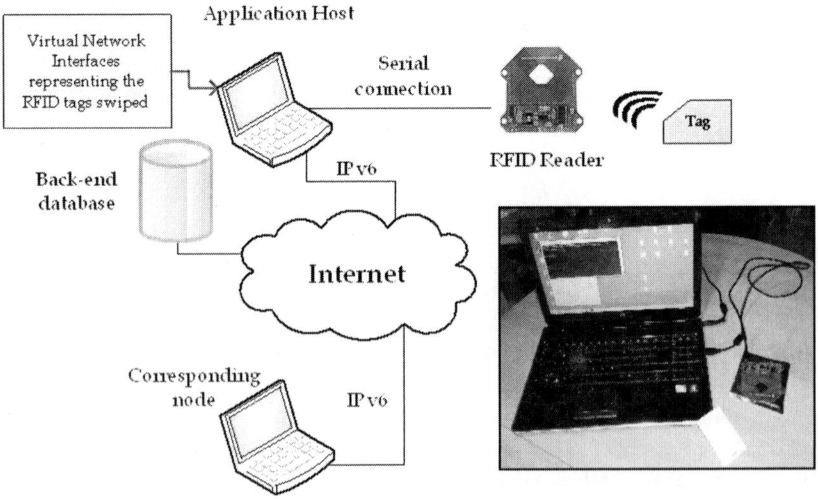

Figure 6. Simple setup to give RFID tags virtual identification on the Internet. A RedBee RFID Reader v1.1 is used. The application is built on the Microsoft®.NET connection software.

Figure 7 shows a state machine diagram for a single VNI resulting from a tag swiped in an access control application.

When the tag is swiped at the reader, the application host creates an IPv6 address by combining the network prefix configured at the reader with the tags identity as illustrated by the Example in Table 1. In the initial state, the software is waiting for a TagSwipe event to occur. Subsequently, the interface is put online with the address constructed, and it is kept alive as long the *expiration time* is greater than 0 (zero) seconds.

Tags are only reachable while they are within reader range. This makes it hard to communicate with the real tag, simply because it is only reachable for a short duration of time. When the tag's attachment to the network is virtualized it is possible to set up an expiration value. This value effectively serves as the time the tags virtual representation on the network can be reached.

The tag identity together with the constructed IPv6 address and a timestamp is stored on the database.Table 1 shows an example of the steps taken to construct an IPv6 address from an EM4100 tag ID.

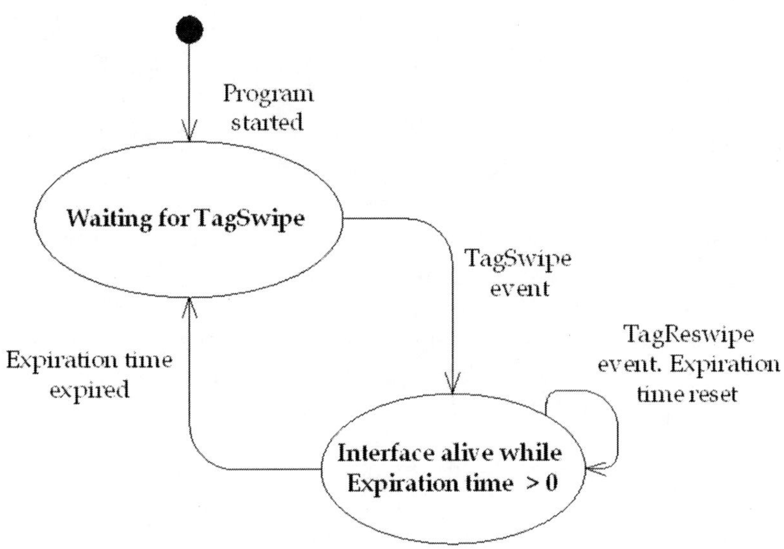

Figure 7. State machine for the virtual interface resulting from a TagSwipe event at the reader.

Table 1. Example of IPv6 network address construction from on EM4100 tag ID.

Tag identity (Example with EM4100 tag)	5 decimal numbers (40 bits)	127 0 58 207 19
Binary ID representation with left-zero-padding	64 bits	0000 0000 0000 0000 0000 0000 0111 1111 0000 0000 0011 1010 1100 1111 0001 0011
Converted to hexadecimal	4 groups of 4 hex. digits	0000 007f 003a cf13
Network prefix of RFID reader (example)	4 groups of 4 hex. digits	2001:16d8:dd92:aaaa::/64
Unicast IPv6 address associated with the tag	8 groups of 4 hex. digits	2001:16d8:dd92:aaaa::00 7f:003a:cf13/128

The application has no visual interface and all configurations must be done in software. For example, it is possible to use more than one virtual interface to represent the tags online. These interfaces need to be preinstalled, as already mentioned, and some parameters in the application need to be configured. Hereafter, it is possible to make use of at least 5 virtual interfaces.

Although focus is on the assignment of IPv6 unicast addresses, tags can also be assigned to become member of multicast groups thereby facilitating one-to-many communication. As an example an application may want to address all tags at a particular reader. Likewise, readers can become members of multicast groups hereby enabling communication to all readers in the multicast group e.g., a particular logical area. Details on how to operate an RFID in IPv6 multicast networks are beyond the scope of this chapter.

6. STRATEGIES FOR INTERWORKING IPV6 WITH RFID TAGS

In this Section we will discuss different methods of mapping between an RFID namespace and an Internet namespace.

6.1. Address Mappings

The most simple approach to find IPv6 addresses for tags is the mapping of the tag ID to an IPv6 address, i.e., the bits from the ID are used to form the IP address [14]. As different passive RFID standards exist there is no common ID structure for tags. Even within standards, there are different types of IDs with different structures. This means, that a general concept to map tag IDs to IPv6 addresses will not work.

Table 2 shows a list of some commonly encountered passive tags and their ID formats.

Table 2. Common passive RFID tag and their characteristics.

IC type	Frequency band	Memory	Standards compliance
EM4100 series	LF (125 KHz)	64 bits	EM4100. Proprietary standard issued by EM Microelectronics [28].

IC type	Frequency band	Memory	Standards compliance
EM4450/4550	LF (125 KHz)	1024 bits	EM4450/4550. Proprietary standard issued by EM Microelectronics [29].
NXP Hitag family (Hitag 1, Hitag 2, Hitag S, Hitag μ)	LF (125 KHz)	256 bit to 2048 bit	IEC/ISO 18000-2 (Hitag μ) [20].
NXP Mifare family (Ultralight/ MF1S20/MF1S50/MF1S70/ DESFire EV1)	HF (13,56 MHz)	64 bytes to 4096 bytes 2K/4K/8K	ISO 14443A [18].
LEGIC Advant family (ATC128, ATC256, ATC1024, ATC2048, ATC4096)	HF (13,56 MHz)	128 bytes to 4096 bytes	IEC/ISO 15693 [19] (ATC128, ATC256, ATC1024), IEC/ISO 14443 A [18] (ATC 2048, ATC 4096).
NXP UCode HSL	UHF (868 MHz or 915 MHz)	2048 bit	ISO18000-4 and 18000-6B [20].
NXP UCode EPC Gen2	UHF (868 MHz or 915 MHz)	512 bit	EPCglobal class 1 gen2 and ISO 18000-6C [20].

An EPC with the length of 64 bits maps well in the IPv6 address format and can result in globally unique addresses. With longer EPCs it is impossible to map the EPC directly into the IPv6 address space and here specialized functions are needed. One solution would be to simply hash the longer EPC's into a length of 64 bits and then use the direct mapping method again. The hashing technique used to derive identifiers was described in Section 3.4, when the CGA namespace was introduced. Another method would be to identify if there are some bits in the longer EPC's that can be removed without affecting the uniqueness property of the tags.

A key benefit of the proposed solution is that there is no need to change the design of existing RFID technology with its EPC namespace conventions. The application can be installed on a computer connected to the reader, and then all objects with RFID tags that pass this reader will put the objects online and thereby giving them the ability to communicate over the Internet as long as the tag is within range of a reader.

Table 3 outlines the different strategies for mapping of tag IDs to IPv6 addresses. Essentially, these divide into methods that work with tags of 64-bit identification or less and tags that use more that 64-bit for identification.

7. MOBILITY CONSIDERATIONS

One of the largest challenges for a dynamic, networked system lies within the mobility support of the network. In the case described here, we consider a system of fixed readers that are connected in a common network infrastructure. Mobility arises when tags are moved between readers. Readers will be wired or wireless and they will have different communication ranges according to their MAC technology. Moreover, they will forward the read tag IDs to the server through the common network infrastructure.

When a tag moves from one reader to another, the network prefix will change but the host suffix/interface ID will still match the tag's EPC. The tag will in effect change its network address every time it passes a new reader. Hence, the challenge is to effectively keep track of tags when the address changes this rapidly.

Table 3. Overview of strategies for mapping Tag ID codes to IPv6 network addresses.

	Strategy/method	Comments
Tag with ID of 64 bits or less	1. Use zero padding left-to-right to create a 64-bit input datum from the tag ID. Map this input to the host suffix/interface ID of the IPv6 address.	The tag ID can be read directly from the IPv6 address as the last 64 bit, i.e., the tag identification works two-ways. There is a risk of address collision with other networked systems that use IPv6 in the same logical subnet e.g., hosts using autoconfiguration. Uniqueness of the RFID naming space in use is conserved.
Tag with ID of more than 64 bits	2. Use CGA namespace adapted to RFID as described in Section 3.4.	Statistical uniqueness is achieved. The tag ID is hashed and cannot be directly read out of the IPv6 address. The reverse process of finding the tag ID from the IPv6 address requires a separate system to perform the mapping since the hashing works only one-way. Extra computational power is required because the method is based on cryptography.
	3. Use the tag ID as input bitstring to create a HIT. The ID should be treated as a public key in accordance with the HIP specification (see Section 3.5). End result is a 128-bit address compliant with the IPv6 addressing format.	Non-compliance with IPv6 address format. Addresses are non-routable because of the separation between locator and the end-point identifier. The solution requires changes in the reader's TCP/IP protocol stack. The solution will function with IPv6 API deployed in applications on the Internet.

There are basically two distinct ways to solve the mobility problem. One is a centralized approach, such as mobile IPv6 [30], where a central server, i.e., the home agent, is used to keep track of the mobile hosts that move around in the world. The mobile IPv6 architecture relies on the concept of a home agent and a care-of address. The method is based on some software on the network layer that can send messages to the home agent making sure that the home agent is holding an updated address list at all times. Initially, traffic destined to the mobile host is routed to the home network and subsequently tunneled to the foreign network that the host is visiting. Fortunately, IPv6 supports mechanisms to circumvent the triangular routing problem that arises in this setup [30].

Dominikus et al. [14], proposed to use mobile IPv6 to handle the mobility of IPv6-enabled tags. In their approach, the care-of address refers to the subnet of the RFID reader, where the tag is currently present. Whilst the care-of address is a globally unique address assigned to the host, i.e., the tag visiting a foreign network, the home agent address is specific to the enterprise using the issued tags.

Alternatively, mobility support can be obtained in a more distributed way by separating location and identity information. This can be achieved by using the HIP approach [22]. In this approach, there is a need to compute the routable IPv6 address from the given non-routable HIT the host has been given.

HIP allows consenting hosts to securely establish and maintain shared IP-layer state, allowing separation of the identifier and locator roles of IP addresses, thereby enabling continuity of communications across IP address changes. A consequence of such a decoupling is that new solutions to network-layer mobility and host multi-homing are possible [22].

8. CONCLUSIONS

Metcalfe's law states that the value of a telecommunications network is proportional to the square of the number of connected users of the system. When the law is applied to a network of objects on the scale predicted by the vision of the Internet of Things it is clear that a single, open architecture for networking physical objects is much more valuable than small scale and fragmented alternatives.

RFID plays an increasingly important in our daily life from management of goods, e-tickets, healthcare, transports, even the identity cards are embedded with RFID tags. In this chapter, we have sketched methods on how to use RFID technology to connect "things" over the Internet by using IPv6. This includes a discussion on the different strategies for mapping of tag IDs to globally unique IPv6 addresses.

For tags with large identification numbers (more than 64 bits) it is proposed to use cryptographic techniques to extract the 64 bits and use these to create a host suffix that is statistically unique.

A testbed used to experiment with the internetworking of low-cost, passive RFID tags to the Internet has been presented. Since these tags do not have electrical and processing power to run an IP protocol stack a virtual network interface (VNI) concept has been introduced. Proxies can be deployed on the edge of the Internet to act on behalf of these passive tags in a protocol message exchange.

To solve the mobility problem, two approaches have been discussed: one being the mobile IPv6 approach and the other being the HIP approach. Both have strengths and both have weaknesses. Mobile IPv6 will need some software to make the connection between the tags and the home agent. The HIP approach needs some computation to take place in order to be able to construct routable IPv6 addresses. Both approaches imply changes to be made to the Internet, as we know it today, before it is possible to effectively achieve the desired results.

Most RFID applications today include mobility as an essential part of their value creation. Therefore, future research in this area must focus on mobility aspects of the Internet of Things.

REFERENCES

1. L. Atzori, A. Iera, and G. Morabito, "The Internet of Things: A survey," Computer Networks, vol. 54, no. 15, pp. 2787-2805, October 2010.

2. S. Sarma, D.L. Brock, and K. Ashton, "The Networked Physical World, Proposals for Engineering the Next Generation of Computing, Commerce & Automatic-Identification," MIT Auto-ID Center Massachusetts Institute of Technology, October 2000. Available from:*http://www.autoidlabs.org/single-view/dir/article/6/93/ page.html*

3. B. Glover and H. Bhatt, "RFID Essentials," O'Reilly, 2006.

4. F. L. Rahman, M.B.I. Reaz, M.A.M. Ali, 2010, "Beyond the Wifi: introducing RFID system using IPv6," Proceedings of ITU-T Kaleidoscobe 2010, pp.1-4. Available from: *http://www.itu.int/ pub/ T-PROC-KALEI-2010*

5. L. Ho, M. Moh, Z. Walker, T. Hamada, and C.-F. Su, "A Prototype on RFID and Sensor Networks for Elder Healthcare: Progress Report," Proceedings of the 2005 ACM SIGCOMM Workshop on Experimental approaches to Wireless Network Design and Analysis (E-WIND'05), pp. 70-75, August 2005.

6. R.H. Jacobsen, Q. Zhang, and T.S. Toftegaard, "Internetworking Objects with RFID," in Deploying RFID - Challenges, Solutions, and Open Issues, Cristina Turcu (Ed.), ISBN: 978-953-307-380-4, InTech. Available from: *http://www.intechopen.com/books/deploying-rfid-challenges-solutions-and-open-issues/internetworking-objects-with-rfid*

7. T. Hornyak, "RFID powder", Scientific American, 298(2), 68-71, February 2008.

8. IDTechEx, "The RFID Knowledgebase." Available from: *http://www.idtechex.com/knowledgebase/en/nologon.asp*

9. J.J. Echevarria, J. Ruiz-de-Garibay, J. Legarda, M. Álvarez, A. Ayerbe, and J.I. Vazquez, "WebTag: Web Browsing into Sensor Tags over NFC," Sensors, vol. 12, pp. 8675-8690, 2012.

10. GS1 EPC Tag Data Standard 1.6, September 2011. Available from: *http://www.gs1.org/gsmp/kc/epcglobal/tds*

11. P. Urien, H. Chabanne, M. Bouet, D.O. De Cunha, V. Guyot, G. Pujolle, P. Paradinas, E. Gressier, and J.-F. Susini, "HIP-based RFID Networking Architecture," IFIP International Conference on Wireless and Optical Communications Networks (WOCN'07), pp. 1-5, July 2007.

12. P. Nikander, J. Laganier, and F. Dupont, "An IPv6 Prefix for Overlay Routable Cryptographic Hash Identifiers (ORCHID)," Internet Society RFC 4843, April 2009. Available from: *http:// tools.ietf.org/html/rfc4843*

13. M. Barisch and A. Matos, "Integrating user identity management systems with the host identity protocol." 2009 IEEE Symposium on Computers and Communications, ISCC, pp. 830-836, 2009.

14. S. Dominikus, M. Aigner, and S. Kraxberger, "Passive RFID Technology for the Internet of Things," 2010 International Conference for Internet Technology and Secured Transactions (ICITST), pp. 1-8, November 2010.

15. R. Moskowitz and P. Nikander, "Host Identity Protocol (HIP) Architecture," Internet Society, RFC 4423. Available from: *http://datatracker.ietf.org/doc/rfc4423/*

16. F. Al-Shraideh, "Host Identity Protocol," International Conference on Networking, International Conference on Systems and International Conference on Mobile Communications and Learning Technologies, ICN/ICONS/MCL 2006., pp. 203, 23-29 April 2006.

17. S. Deering and R. Hinden, "Internet Protocol, Version 6 (IPv6) Specification," Internet Society, RFC 2460, December 1998. Available from:*http://tools.ietf.org/html/rfc2460*

18. "ISO/IEC 14443 Identification cards - Contactless integrated circuit(s) cards - Proximity cards." Available from: *http://wg8.de/sd1.html#14443*

19. "ISO/IEC 15693 Identification cards - Contactless integrated circuit(s) cards - Vicinity cards." Available from: *http://wg8.de/sd1.html#15693*

20. "ISO/IEC 18000 Information technology - Radio frequency identification for item management." Available from:*http://www.iso.org/iso/home/store.htm*

21. Bose and R. Pal, "Auto-ID: managing anything, anywhere, anytime in the supply chain," Communications of the ACM, vol. 48, no. 8, pp. 100-106, August 2005.

22. P. Nikander, T. Henderson, C. Vogt, and J. Arkko, "End-Host Mobility and Multihoming with the Host Identity Protocol", Internet Society, RFC 5206, April 2008. Available from: *http://tools.ietf. org/html/rfc5206*

23. D. Eastlake and P. Jones, "US Secure Hash Algorithm 1 (SHA1)," Internet Society, RFC 3174, September 2001. Available from: http://tools.ietf.org/html/rfc3174

24. R. Rivest, "The MD5 Message-Digest Algorithm," Internet Society, RFC 1321, April 1992. Available from: http://tools.ietf.org/ html/rfc1321

25. B. Xu, Y. Liu, X. He, and Y. Tao, "On the Architecture and Address Mapping Mechanism of IoT," 2010 International Conference on Intelligent Systems and Knowledge Engineering (ISKE), pp. 678-682, November 2010.

26. Pappas, S. Hailes, and R. Giaffreda, "Mobile Host Location Tracking through DNS," London Communications Symposium (LCS) and

Photonics London, LCS 2002, Available from: *http://www.ee.ucl.ac.uk/lcs/previous/LCS2002/lcs2002.html*

27. T. Aura, "Cryptographically Generated Addresses (CGA)," Internet Society, RFC 3972, March 2005. Available from:*http://tools.ietf.org/html/rfc3972*

28. EM Microelectronics, "EM4100 / Read Only Contactless Identification Device," data sheet, 2004. Available from*http://www.datasheetarchive.com/*

29. EM Microelectronics, "EM4450/4550 / 1 KBit Read/Write Contactless Identification Device," data sheet, 2010. Available from:*http://www.datasheetarchive.com/*

30. D. Johnson, C. Perkins, and J. Arkko, "Mobility Support in IPv6," Internet Society, RFC 3775, June 2004. Available from:*http://tools.ietf.org/html/rfc3775*

7

UNDERSTANDING RFID TECHNOLOGY WITHIN A BUSINESS INTELLIGENCE FRAMEWORK

Richard Herschel, Patricia D. Rafferty

Erivan K. Haub School of Business, Saint Joseph's University, Philadelphia, USA

ABSTRACT

Radio Frequency Identification (RFID) technology provides new and exciting opportunities for increasing organizational, financial, and operational performance. With its focus on organizational efficiency and effectiveness, RFID technology is superior to barcodes in its ability to provide source automation features that increase the speed and volume of data collection for analysis. Today, applications that employ RFID are growing rapidly and this technology is in a continuous state of evolution and growth. As it continues to progress, RFID provides us with new opportunities to use business intelligence (BI) to monitor organizational operations and learn more about markets, as well as consumer attitudes, behaviors, and product preferences. This technology can even be used to prevent potentially faulty or spoiled products from ending up in the hands of consumers. However, RFID offers significant challenges to organizations that attempt to employ this technology. Most significantly, there exists the potential for RFID to overwhelm data collection and BI analytic efforts if organizations fail to effectively address RFID data integration issues. To this end, the purpose of this article is to explicate the dynamic technology of RFID and how it is being used today. Additionally, this article will provide insights into how RFID technology is evolving and how this technology relates to BI and issues related to data integration. This knowledge has never been more essential. While IT academic research into RFID development and issues has declined in recent years, RFID continues to be a vital area of exploration, especially as it relates to BI in the 21st century.

KEYWORDS

Business Intelligence; Radio Frequency Identification Technology; Organizational Performance; Data Integration; Data Analytics; Supply Chain Technology; Barcode Data

1. INTRODUCTION

There has been substantial academic research conducted on Radio Frequency Identification (RFID) technology, along with its capacity to enhance organizational decision-making. Some of this research has focused on RFID and information systems; however, this has diminished as a topic of examination over recent years. This is unfortunate and ill timed, as there have recently been many noteworthy developments and issues that warrant scrutiny, especially with regard to the relationship between RFID and business intelligence.

RFID has been identified as one of the most important technologies of the 21st century because it allows for real-time tracking of objects or products within the supply chain [1]. RFID is a wireless technology that incurporates a tag containing a chip and an antenna for sending and receiving data. The technology interfaces with an interrogator (reader), which has an antenna that communicates with the tag. Middleware then manages, filters, aggregates, and routes the data captured from the tag to the organization's enterprise systems for storages and analyses [2]. This type of technology offers an improved alternative to barcodes. For example, scanning barcodes requires labor, which is costly. Because it is more expensive to collect information via barcodes, retail companies typically conduct inventory counts only once or twice a year, often yielding inaccurate data despite these costly efforts. However, inventory accuracy for apparel stores averages just 65 percent. Unfortunately, this problem just scratches the surface. Retailers consistently have a lack of meaningful and high-volume data regarding the movement of goods from their warehouse to the back of the store, then on to store shelves. Barcodes are not typically scanned when items arrive at the warehouse, nor when they are shipped to a store, arrive at that store, when the items are placed at a particular location within the confines of the store, including the store's selling area. As a result, retailers lack the information and knowledge of how long these processes take to complete or how well the stores are executing these tasks [3].

Similarly, barcode data cannot provide information as to whether an item did not sell because it was unpopular with shoppers or because it was simply missing from the shelves when shoppers were ready to buy. This inability to accurately measure product sales success is one of the most fundamental questions for retail operations today; however, barcode technology does not address this essential question, thus hindering meaningful merchandiseing improvements.

These impediments are readily overcome by RFID technology because, unlike barcodes, RFID is not constrained by a need for line-of-sight reading and multiple RFID tags can be read simultaneously. Similarly, RFID tags can store critical information regarding the individual item, can be read in a dirty environment, and have the potential as reusable data containers since they have the potential to be written multiple times. These resultant advantages to barcodes have made RFID a critical component to an enterprise's data supply chain. RFID automates data collection for goods and services, which allows for the application of both automated and ad hoc analytics. This drives drive decision-making capacity that seeks to enhance organizational and process-oriented effectiveness and efficiency [3].

This organizational and process-oriented effectiveness and efficiency is most apparent in RFID technology's capacity to function on a proximity basis. RFID tags can be read and processed as long as they pass within a given radius of a RFID reader. Unlike barcode technologies, where each barcode tag must pass directly in front of a reader, hundreds of RFID tags can be read simultaneously, since there is no requirement for line-of-sight scanning of each tag. In shipping applications, this enables scanning and tracking to be done at a palette level, with tagged items remaining snuggly packaged in a box or shipping crate. RFID eliminates the need to scan each tag on a one-by-one basis. This ease and enhanced speed of data collection affords firms the opportunity to use BI more rapidly, as well as exploit or correct situations that have a direct impact on enterprise performance.

2. THE EVOLUTION OF RFID

Predominantly used in the consumer goods industry [4], RFID technology continues to evolve. Today's enterprises now employ RFID data for many diverse applications, including airline luggage tracking, electronic security, asset tracking, and supply-chain management [5]. Other uses for RFID include monitoring out-of-stock merchandise, product distribution, receiving documentation, proof of delivery, asset management, product life cycle tracking, recall management, product rotation, inventory and shipping accuracy, inventory management, product shrinkage, military applications, and service sector applications [6-9]. Facebook is experimenting with RFID chips embedded in wristbands or badges that allows its members to carry out a variety of social networking tasks without the need for a computer or smartphone [10].

Many governments and academic institutions now utilize RFID tags. For example, RFID technology is being adopted for such applications such as tracking farm animals, non-stop road tolling, library and document tagging, and national ID cards and passports. Colleges and universities are beginning to implement RFID identification for attendance and alert solutions. For example, Scholar Chip, a New York-based provider of RFID attendance and payment solutions, has issued close to 750,000 RFID tags in various formats, such as ID cards, fobs and stickers. Students hold their RFID-enabled ID badges to readers as they enter the school building, and again as they enter classrooms, where readers are mounted on teachers' desks. Additionally, Universidad Regiomontana in Mexico is using RFID technology to prevent theft of laptop computers issued to staff and faculty [11].

Further examples of loss/theft protection is apparent in the manner in which RFID tags are poised to provide anti-theft for laptops and tablet computers using a Wi-Fi tag. Ekahau, a provider of Wi-Fi-based real-time location system solutions, has announced the launch of a Wi-Fi tag that incorporates a location-tracking device designed to prevent laptop theft and misplacement. The RFID embedded device is a small, flat unit that is plugged into an ExpressCard and can be activated via a few keystrokes. The tag is designed to track and manage the tags anywhere on a campus equipped with Wi-Fi coverage. By determining a tag's location and status, the system can alert a user if a laptop or tablet computer leaves a designated area, or enters an unsecured area [12].

2.1. RFID and the IT Organization

IT organizations are also becoming a major consumer of RFID technology. Market revenue for RFID tags, hardware (such as readers and printers), software and middleware, and integration services used to track and trace IT assets within data centers reached $96.3 million in 2011. Revenue is expected to grow to $952.6 million in 2017, due to the widespread implementation of RFID technology in data centers globally. RFID tags can quickly locate and track innumerable servers and vast numbers of data tapes within large data centers, thus lowering their replacement expenses. Likewise, the lack of asset visibility sometimes results in the duplication of existing assets, ultimately adding to total costs. RFID technology can eliminate this duplication, thus providing significant savings, as well as preventing theft and loss through real-time tracking [13].

RFID provides opportunities for organizations to realize cost efficiencies and effectiveness as they seek to streamline their physical processes. Roberti [14] suggests five ways RFID technology facilitates these objectives:

- Deploying RFID to track the 7 - 10 percent of shipping containers typically replaced annually because they are lost or damaged;

- Tagging IT assets with RFID tags instead of bar codes enables more efficiency for tracking and accounting;

- Providing suppliers with RFID tags to be placed on shipments so that the manufacturer can more efficiently and more accurately associate them with parts, raw materials, or products;

- Using RFID to track rental equipment utilization;

- Deploying RFID on subassemblies, large parts, and parts bins to increase throughput with less labor and costs associated with locating parts and subassemblies.

2.2. RFID Growth Potential

The number of RFID applications continues to present unbounded opportunities, with diverse industries applying this technology throughout the world. In 2006, cumulative sales of RFID tags reached

$2.4 billion (USD) [15]. In 2011, the value of the entire RFID market was $5.84 billion (USD), up from $5.63 billion (USD) in 2010 [16]. RFID market value continued to grow in early 2012, where the cumulative number of RFID tags sold was 15.1 billion, 20 percent of which were sold last year [13].

Even with this unprecedented growth, the RFID market is predicted to grow steadily over the next decade, rising four-fold within that period to $26.19 billion (UDD) in 2022. A recent research report entitled "Global RFID Market Forecast to 2014" states that the RFID market is expected to grow at a compound annual growth rate of around 18% through 2014 to reach approximately USD 19.3 billion. It also notes that the RFID market in the Asia-Pacific is growing at a faster rate than those of Europe and America [17]. Further, according to ABI Research, it is expected that $70.5 billion will be generated between 2012 and 2017 from sales of RFID readers, transponders, software and services, with sixty percent of the accumulated revenue over the next five years being generated from government, retail, transportation and logistics. These four sectors have been identified by ABI as the most valuable sectors for RFID growth potential. The market is expected to grow 20 percent, year over year, per annum [18].

To foster the further growth in the deployment of this technology, new and cheaper RFID chips are now under development. For example, microelectronics developer, Terepac, reports that it plans to begin manufacturing a "near field communication" (NFC) RFID tag that is expected to be the world's smallest RFID tag. This tag would be cheaper and much smaller than other NFC tags currently on the market, and thus could be embedded in almost any paper label, product or item. The tag could be interrogated via an NFC reader in a user's mobile phone, thereby making it possible to link a customer with information regarding a particular product, or link an individual with a social-networking site, medication data or other Web-based information [19].

2.3. The Future of RFID Technology

At MIT, Bhattacharyya, Di Leo, Floerkemeier, Sarma and Anand [20] are working on an inexpensive, passive RFID chip that may dramatically enhance data collection on the status of food items requiring refrigeration. Their research reports on the design of a temperature RFID threshold sensor that registers the violation of critical temperature thresholds by inducing a permanent change in an RFID tag antenna's electrical properties. This inexpensive sensing tag would be useful in monitoring the state of certain items, such as perishable food produce or other fragile goods that must maintain a constant temperature threshold at all times. The sensor would provide the RFID user with binary state information indicating whether critical temperature thresholds were reached during transportation allowing the user to make a decision of whether to keep or discard the goods. Since the state of most goods are examined at the distribution centers in a supply chain, rather than in transit, this type of RFID advancement would dramatically improve realtime monitoring for items, such as orange juice cartons, milk and meat products. Within supply chain applications, an accurate and timely assessment of the state of the product will be automatically monitored as soon as the products pass an RFID reader at the dock door of a distribution center.

These types of examples illustrate the manner in which RFID technology continues to evolve and innovate at an unprecedented rate, not only in the design of the technology, but also in its application. In 2011, the RFID journal reported 101 innovative ways to use RFID [21]. New RFID applications include saving endangered species, smart supermarket carts, liquor pouring tracking at bars, protecting miners, cigarette smoking cessation programs, at-home pain management, and the detection of gunfire.

3. RFID TECHNOLOGY AND ACADEMIC RESEARCH

RFID technology has been an exciting area for academic study due to its relative novelty and exploding growth [22]. Scholarly and practitioner-based exploration into RFID technology has led to the emergence of a new research area that builds on existing study in a host of diverse disciplines, such as electronic engineering, information systems, computer science, and business strategy. This has resulted in a significant increase in the number of scholarly papers on RFID in research journals from 1995 to 2005. During this time, topical areas of research provided a general view of the use of RFID technology in the areas of retail and supply chains. This research also dealt with the potential of utilizing RFID technology, the perceived benefits, effects, and challenges for retailers, how consumers are likely to respond to the technology, and the market drivers for RFID implementation in the grocery industry. Within this 10-year period, the majority of research had been published in technical journals, but it is expected that papers related to RFID technology will continue to grow in management and other business journals.

Bendavid, Lefebvre, Lefebver, & Fosso-Wamba [4] note that RFID has been a topic of interest in various fields of research related to the technology itself, such as: 1) Innovation management and potential trajectories for RFID adoption; 2) Implementation in an inter-organizational context; 3) Customer relationship management (CRM); 4) Product lifecycle management (PLM); 5) Reverse supply chain activities (e.g., returned goods); and 6) Supply chain management. Recent research has also investigated the impact of RFID technology on retail sales and service operations. For example De Marco, Caglian, De Marco, Caglian, Nervo, and Rafele [23] found that sales growth resulted from the dynamic and integrated impacts of RFID technology, especially in the areas of improved inventory control, faster inventory turnover, and longer time available for store personnel to assist consumers as an effect of more efficient backroom operations.

Ferrer, Dew, and Apte [9] studied several different RFID applications within the service sector to discover how this technology can be used to enhance various aspects of the service delivery, as well as the costs and benefits realized form this technology. Their research reveals how RFID impacts service operations in a variety of scenarios. The authors note that decisions about the selection of new technologies, such as RFID, require significant investment and can have a strategic impact on the organization. To ensure that the right RFID configuration is selected, Ferrer et al. assert that management must be an informed consumer of the technology in order to understand how the data adds value to their operations. In their investigation of RFID utilization in service industries, Ferrer et al. identify four benefits that RFID applications provide, including: 1) Replacement of labor through automation; 2) Cycle time reductions; 3) Enabling self-service; and 4) Advances in loss prevention. They argue that these benefits lead to better control and lower costs.

Baars, Kemper, Lasi, and Siegel [24] find that the dominant focus of RFID research today deals with increasing local efficiency gains by implementing a more efficient handling of items or pallets, along with a reduction of processing times. However, while substantial benefits are expected to result from the sharing and processing of RFID data along the supply chain, Baars et al. argue that these applications do not yet take the full potential of RFID data into account. They argue that RFID also requires integrating, enriching and analyzing RFID data for the purpose of supporting decision makers. Supporting data-driven decision making means paying special attention to how RFID technology affects and supports the use of business intelligence capabilities within an organization.

4. BI AND RFID TECHNOLOGY

BI is widely adopted in organizations and successful BI initiatives have been undertaken across major industries and for varied applications, including health care [25,26], security and event management [27], telecommunications [28], web analytics [29], and Six Sigma [30].

Herschel [31] defines BI as the application of data, technology, and analytics in the pursuit of insights and knowledge that enables decisions and actions that yield value for a firm. He asserts that BI creates value by providing evidence that organizations can use to make informed decisions about people, processes, products and services. Herschel's definition suggests that because RFID technology has inherent capabilities for capturing and transmitting product data for decision-making, it should be seen as an important data source for BI.

Figure 1 presents a vision of BI as an integrative application of technologies, models, techniques, and practices. In Miori and Klimberg's [32] framework, each of the three circles of the Venn diagram represent applications that had previously been considered quite distinct, which include 1) information systems and technology, 2) statistics, and 3) operation research/management science.

This framework serves to encapsulate a broad definition of BI, where BI is characterized from each of three viewpoints, which consist of 1) business information intelligence (BII), 2) business statistical intelligence (BSI) and 3) business modeling intelligence (BMI). Each of these three viewpoints has particular business aspects that are independent of the other viewpoints. Conversely, each viewpoint can work together or utilize techniques/skills from one or possibly two of the other disciplines. For example, data mining, which requires a high level of statistical knowledge as well as the availability of necessary data, may require significant IT skills and/or knowledge. Further, if data mining analysis demands a systematic process of analysis, modeling skills may be required. Within the context of this framework, data provided by RFID would be secured by IT technology and then be subjected to statistics and modeling analytics to derive understanding of the data so as to affect decision-making.

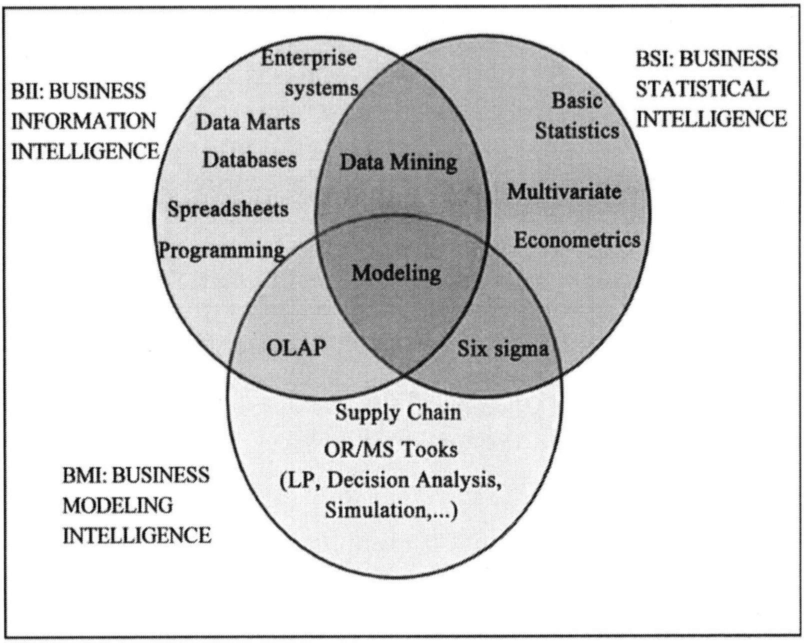

Figure 1. Business intelligence/business analytics breakdown (Klimberg & Miori, 2010).

4.1. RFID Technology and Decision-Making

The confluence of business intelligence and RFID technology is best illustrated in the manner in which BI provides the means by which RFID data can influence business processes and decision-making. BI also provides the underlying architectures, tools, databases, applications, and methodologies that support RFID technology and its ability to positively affect organizational processes [33]. Within this context, the major objectives of BI are to enable interactive and easy access to diverse data, to enable manipulation and transformation of this data, and to provide business managers and analysts with the ability to conduct appropriate analyses for actionable execution [28]. BI is seen as a data-centric approach that relies on advanced data collection, extraction, and analysis technologies [34].

Since data warehousing is the foundation of BI, the design of data marts and tools for extraction, transformation, and load (ETL) are essential for converting and integrating enterprise-specific data. Database query, online analytical processing (OLAP), and advanced reporting tools are typically adopted to explore important data characteristics. Business performance management (BPM) tools, such as scorecards and dashboards, are often used in BI to analyze and visualize various performance metrics. In addition to the business analytics functions inherent in BI software, advanced knowledge discovery using data and text mining are usually adopted for association rule mining, database segmentation and clustering, anomaly detection, and predictive modeling in various business applications. BI technologies provide historical, current, and predictive views of business operations to enhance the comprehension/understanding of fact-based interrelationships and this makes BI an enabling technology for knowledge creation [35].

Within the context of BI, the better the data, the more accurate are the results of the analytics that employ the data. This in turn leads to more informed and (hopefully) better decision-making [36]. In BI, the ideal means for capturing data is via source automation, since this process increases both the efficiency and the accuracy of data entry and it is cost effective because it eliminates much of the effort and delay caused when data must be transcribed. RFID technology is a source automation vehicle since it provides an automated, efficient, cost-effective means of providing accurate data for BI evidence-based decision making. RFID data is used to build up an “internet of things”, which can be viewed as a network that allows companies to track goods through the global supply chain and execute many applications simultaneously [37]. In addition, RFID source automation capabilities deliver data that BI can use to prevent potential economic damage caused by product piracy. This is possible because RFID tag data is effectively used for product authentication in the supply chain [38].

4.2. Affiliated Costs

However, both data quality and availability depend, in part, upon the cost of acquiring the data. Much of the cost for data is derived from getting the data out of source systems (extract), and transforming it to make it consistent across all the various dimensions (transform), and then putting it in a model that is easy to query and analyze (load). Data accuracy inevitably affects task accuracy and the time spent solving a problem [39].

Consequently, there is a cost for placing RFID tags on products to capture data. The RFID Journal [40] notes that most companies that sell RFID tags do not quote prices because pricing is based on: 1) Volume; 2) the amount of memory on the tag; 3) the packaging of the tag, for instance whether it's encased in plastic or embedded in a label; 4) whether the tag is active or passive, and other dimensions. In general, active tags are priced at $25 and up. An active tag is equipped with a battery that can be used as a partial or complete source of power for the tag's circuitry and antenna. However, active tags with special protective housing, extra-long battery life, or sensors can cost $100 or more. A passive tag is an RFID tag that does not contain a battery and relies on a tag reader that supplies the power. The price for these chips range from seven to 15 US cents. However, the RFID Journal notes that if the passive tag is embedded in a thermal transfer label on which companies can print a bar code, the price rises to 15 cents and up.

5. BI AND DATA INTEGRATION

Within the context of BI, RFID technology provides a plethora of new data for collection and analysis. However, Edwards [41] asserts that integrating RFID-generated data into BI and other business applications so that essential information can be viewed, analyzed, applied, and presented to end users is often more difficult than assembling the data. Since RFID is used in diverse ways by a variety of organizations and with many different applications, it is problematic to create a single, all-purpose RFID data-integration strategy. This leads many businesses to develop unique integration concepts and tools, or to adopt an amalgamation of approaches used by other organizations that end up thwarting analytic efforts.

In order to enable a more effective relationship between BI and RFID technology, organizations will need to pay more attention to RFID and data integration. To accomplish this, firms will need to better understand RFID technologies and practices, how the technology can be used to improve business processes, and better understand the nature of RFID data. For example, the use of RFID often requires the serialization of products or other assets, which means that RFID adopters need to give individual items a unique serial number, instead of describing them by name or by using randomly allocated numbers. Also RFID enabled real-time location data is often difficult for organizations to integrate into existing ERP applications than other types of RFID information.

Consequently, full utilization of RFID technology would require significant business process redesign and the associated development of new and retooled information technology capabilities within an organization [42]. From an information systems perspective, RFID could be viewed as another technology for capturing source data. Although RFID technology holds the promise of process innovations that create significant improvements in supply chain business models, organizations would still face a new, longer-range set of challenges to create significant economic value. These challenges involve the development of information systems solutions that integrate data derived from RFID into business processes; and developing the feedback-loops, such that new sources of information generate more efficient and effective business processes and decision-making.

The potential for RFID to overwhelm data collection and analytic efforts is significant, since the technology continues to evolve and its applications rapidly expand. This means RFID will contribute to the increasing volume and detail of information captured by enterprises. For instance, IBM [43] states that 2.5 quintillion bytes of data are created each day—so much that 90% of the data in the world today has been created in the last two years alone. This data comes from a plethora of sources and is oftentimes referred to as "big data".

Hence, while RFID provides an important mechanism for fostering new BI applications with a wider range of timeand location-informing data, firms need to better understand RFID technology itself, the data integration issues RFID can create, and decide which BI efforts they want RFID data to support.

6. OPPORTUNITIES FOR RESEARCH

In an era of economic uncertainty, many organizations, both public and private, have looked at options to reduce costs. Many have reduced their workforce and/or outsourced work, which also includes the trend for organizations to move toward cloud computing to reduce their IT costs. RFID technology has the potential to enable firms to further reduce their costs, gain operational efficiencies, and to learn more about their customer behavior and processes.

However, the full potential of RFID technology has yet to be realized, primarily because many organizations are still dependent on bar codes in their source automation applications. Early in the millennium, research institutions and universities forecasted a revolution in identification of goods within a few years, but this never happen. The failure to transition to RFID was not due to the cost of the tags, but rather that retailers had made substantial investments in barcode technology and weren't willing to change just because the new technology was available.

To address the RFID adoption lag, Impinj, a provider of RFID solutions for identifying, locating, and authenticating items, has unveiled a new Store Performance Simulator (SPS). This Web-based analysis tool is designed to enable retailers considering the deployment of RFID technology to perform what-if scenarios. The tool will help retailers better understand if RFID can increase inventory accuracy and improve a store's profitability. The SPS can help retailers begin to understand the impact, not just of RFID technology, but also of adding additional service labor or inventory-management labor, or of shortening their supply chain. Instead of spending tens of thousands of dollars to test results in the real world, organizations can simply perform a virtual simulation. SPS will provide BI capabilities to retailers, allowing them to forecast the impact of using RFID technology on operating profit, inventory accuracy, unit sales and other metrics [44].

In the future, it is expected that RFID will be able to collect sales information by location and fixture within a store. This will make it feasible to test how well an item sold at one location versus another at a different store, and to optimize a store's configuration in order to maximize revenue. Retailers do some of this now, but their store configurations are often based on selling price, or on mere guesswork. With hard data about an item's location and sale price, the amount of time that product spends on the floor, and other data that an item-level RFID system can supply, retailers will be able to simulate different store configurations and forecast sales. This, in turn, will provide better BI input and enable retailers to make better, more informed decisions.

Within the BI context, having real data regarding shipping accuracy, replenishment times, pick accuracy, etc., allows companies to measure—and improve—their performance. Moreover, due to the frequency that RFID data can be collected, more efficient BI allows retailers to correct mistakes more rapidly. However, one impedement for the adoption and deployment of RFID technology has been the inconsistent development and adoption of universal standards. There are accepted standards for many facets of RFID, including near-field communication for payment cards and many industrial applications. However, there are still applications where the RFID standards are not uniform. To address this deficiency, the International Standards Organization (ISO) is taking a leading role in developing conclusive, comprehensive RFID standards across the board.

Finally, it must be noted that much of what has been reported in academic information systems journals concerning RFID technology is dated. As RFID growth continues to accelerate, it is clear that there are noteworthy and substantial prospects for academic research. Clearly, there will be ongoing occasions to track and monitor both the successes and failures that occur in RFID development, implementation and application. In addition, there will be substantial opportunities to better understand how RFID technology and BI efforts combine to influence, enhance, or encumber organizational decision-making.

REFERENCES

1. C. C. Chao, J. M. Yang and W. Y. Jen, "Determining Technology Trends and Forecasts of RFID by Historical Review and Bibliometric Analysis from 1991 to 2005," Technovation, Vol. 27, No. 5, 2007, pp. 268-279.

2. J. Curtin, K. Kauffman and F. J., Riggins, "Making the 'MOST' out of RFID Technology: A Research Agenda for the Study of the Adoption, Usage, and Impact of RFID," Information Technology and Management, Vol. 8, No. 2, 2007, pp. 87-110

3. C. Riemenschneider, W. Hardgrave and D. Armstrong, "Is There a Business Case for RFID?" Working Paper ITRI-WP091-0507, University of Arkansas, Fayetteville, 2007.

4. U. Bendavid, E. Lefebvre, L. Lefebvre and S. FossoWamba, "Key Performance Indicators for the Evaluation of RFID-enabled B-B E-Commerce Applications: The Case of a Five-layer Supply Chain," Information Systems E-Business Management, Vol. 7, No. 1, 2009, pp. 1-20.

5. S. F. Tzeng, W. H. Chen and F. Y. Pai, "Evaluating the Business Value of RFID: Evidence from Five Case Studies," International Journal of Production Economics, Vol. 112, No. 2, 2008, pp. 601-613.

6. G. Q. Huang, J. S. K. Lau and K. L. Mak, "The Impacts of Sharing Production Information on Supply Chain Dynamics: A Review of the Literature," International Journal of Production Research, Vol. 41, No. 7, 2003, pp. 1483-1517.

7. L. Lapide, "RFID: What's in It for the Forecaster?" The Journal of Business Forecasting Methods & Systems, Vol. 23, No. 2, 2004, pp. 16-20.

8. B. Oztaysi, S. Baysan and F. Akpinar, "Radio Frequency Identification in Hospitality," Technovation, Vol. 29, No. 9, 2009, pp. 618-624.

9. G. Ferrer, N. Dew and U. Apte, "When is RFID Right for Your Service?" International Journal of Production Research, Vol. 124, No. 2, 2010, pp. 414-425.

10. N. Clayton, "Facebook Finds Ways to Shift Social Networking Online," 2011. http://blogs.wsj.com/tech-europe/2011/08/05/facebook-finds-ways-shift-social-networking-offline/?KEYWORDS=rfid

11. M. S. O'Connor, "RFID Goes to School," 2012. http://www.rfidjournal.com/magazine/article/9577

12. RFID Journal, "RFID News Roundup," 2012. http://www.rfidjournal.com/article/view/9692

13. RFID Journal, "RFID News Roundup," 2012. http://www.rfidjournal.com/article/articleview/9746/1/1

14. M. Roberti, "Global Economic Uncertainty and Slower Growth," RFID Journal, 2011, pp. 6-8.

15. R. Das, "RFID Tags in 2005—How Many and Where," IDTechEx, New York, 2005

16. R. Das and P. Harrop, "RFID Forecasts, Players and Opportunities 2011-2021," 2011. http://www.idtechex.com/research/reports/rfid_forecasts_players_and_opportunities_2011_2021_000250.asp

17. RFID World, "ABI Research: RFID Market to Surpass $70 Billion over Next Five Years," 2012. http://www.rfidworld.ca/abi-research-rfid-market-to-surpass-70-billion-over-next-five-years/860

18. RFID World, "World RFID Market to Reach 20 billion USD in 2014?" 2012. http://www.rfidworld.ca/world-rfid-market-to-reach-20-billion-usd-in-2014/769

19. C. Swedberg, "Terepac Announces Tiny NFC RFID Tags," 2012. http://www.rfidjournal.com/article/view/9661

20. R. Bhattacharyya, C. Di Leo, D. Floerkemeier, S. Sarma and L. Anand, "RFID Tag Antenna Based Temperature Sensing Using Shape Memory Polymer Actuation," IEEE Sensors Conference, Kona, 1-4 November 2010, pp. 2363- 2368.

21. "RFID for What? 101 Innovative Ways to Use RFID," RFID Journal, 2011, pp. 11-35.

22. E. Ngai, K. Moon, F. Riggins and C. Yi, "RFID Research: An Academic Literature Review (1995-2005) and Future Research Directions," International Journal of Production Economics, Vol. 112, No. 2, 2008, pp. 510-520.

23. A. De Marco, A. Cagliano, M. Nervo and C. Rafele, "Using System Dynamics to Assess the Impact of RFID Technology on Retail Operations," International Journal of Production Economics, Vol. 32, No. 1, 2012, pp. 333- 344.

24. H. Baars, H. G. Kemper, H. Lasi and M. Siegel, "Combining RFID Technology and Business Intelligence for Supply Chain Optimization: Scenarios for Retail Logistics," Proceedings of the 41st Hawaii International Conferenceon System Sciences, Waikoloa, 7-10 January 2008, pp. 1-10.

25. T. A. Carte, A. B. Schwarzkopf, T. M. Shaft and R. W. Zmud, "Advanced Business Intelligence at Cardinal Health," MIS Quarterly Executive, Vol. 4, No. 4, 2005, pp. 413- 424.

26. A. Olinsky and P. Schumacher, "Data Mining for Health Care Professionals: MBA Course Projects Resulting in Hospital Improvements," International Journal of Business Intelligence Research, Vol. 1, No. 2, 2010, pp. 30-41.

27. K. Lozito, "Mitigating Risk: Analysis of Security Information and Event Management," International Journal of Business Intelligence Research, Vol. 2, No. 2, 2011, pp. 67-75.

28. E. Turban, S. Sharda, J. E. Aronson and D. King, "Business Intelligence: A Managerial Approach," Pearson/Prentice Hall, Upper Saddle River, 2008.

29. L. Iyer and R. Raman, "Intelligent Analytics: Integrating Business Intelligence and Web Analytics," International Journal of Business Intelligence Research, Vol. 2, No. 1, 2011, pp. 31-45.

30. D. Miller, "Improving Business Intelligence: The Six Sigma Way," International Journal of Business Intelligence Research, Vol. 2, No. 1, 2010, pp. 31-45.

31. R. T. Herschel, "Business Intelligence: Opportunities for Research," Proceedings of the 33rd International Conferenceon Information Technology Interfaces, Cavtat, 27- 30 June 2011, pp. 1-5.

32. R. Klimberg and V. Miori, "Back in Business: OR/MS Today," 2010. http://www.informs.org/ORMS-Today/Public-Articles/October-Volume-37-Number5/Back-in-Business

33. M. Raisinghani, "Business Intelligence in the Digital Economy: Opportunities, Limitations and Risks," The Idea Group, Hershey, 2004.

34. H. J. Watson and B. Wixom, "The Current State of Business Intelligence," IEEE Computer, Vol. 40, No. 9, 2007, pp. 96-99.

35. D. Steiger, "Decision Support as Knowledge Creation: A Business Intelligence Design Theory," International Journal of Business Intelligence Research, Vol. 1, No. 1, 2010, pp. 29-47.

36. T. Davenport, "Competing on Analytics," Harvard Business Review, Vol. 84, No. 1, 2006, pp. 98-107.

37. B. Violino, "Leveraging the Internet of Things," RFID Journal, 2005, pp. 1-2.

38. M. Lehtonen, F. Michahelles and E. Fleisch, "Trust andSecurity in RFID-based Product Authentication Systems," Systems Journal, Vol. 1, No. 2, 2007, pp. 129-144.

39. D. Biros, J. George and R. Zmud, "Inducing Sensitivity to Deception in Order to Improve Decision Making Performance: A Field Study," MIS Quarterly, Vol. 26, No. 2, 2002, pp. 119-144.

40. RFID Journal, "How Much Does an RFID Tag Cost Today?" 2011. http://www.rfidjournal.com/faq/20/85

41. J. Edwards, "Integrating RFID Data with Business Apps: Seven Strategies that Work," RFID Journal, 2012, pp. 14-23.

42. F. Neiderman, R. Mattieu, R. Morley and L. W. Kwon, "Examining RFID Applications in Supply Chain Management," Communications of the ACM, Vol. 50, No. 7, 2007, pp. 93-101.

43. IBM, "Brining Big Data to the Enterprise, 2012. http://www-01.ibm.com/software/data/bigdata

44. M. Roberti, "RFID and Data-Driven Retailing," 2012. http://www.rfidjournal.com/article/articleview/9621/1/2

8

DESIGN AND IMPLEMENTATION OF RFID-BASED OBJECT LOCATORS

T. S. Chou[1] and J. W. S. Liu[2]

[1] Computer Science Department, University of California at Irvine, Irvine, CA, USA

[2] Institute of Information Science, Academia Sinica, Nangang, Taipei, Taiwan

1. INTRODUCTION

In the coming decades, an increasingly larger number of baby boomers will grow into old age. This trend has led to an increasing demand for devices and services (e.g., [1-8]) that can help elderly individuals to live well and independently. *Object locator* is such a device. The device can assist its users in finding misplaced household and personal objects in a home or office. Figure 1 shows several object locators offered today by specialty stores and websites. Each of these locators contains an interrogator with a few buttons and an equal number of tags: Even the largest one, the leftmost one in the figure, offers only 8 buttons. The buttons are of different colors, and there is a tag of the color matching the color of each button. By attaching a tag to an object to be tracked, the user can look for the object by pressing the button of matching color on the interrogator. The tag attached to the object beeps and flashes in response and thus enables the user to find the object. Other locators work similarly.

Existing object locators are not ideal in many aspects: The number of buttons on the interrogator and tags is fixed, and the number is small. Extending the locator to track more objects is impossible. – If the user were to use more than one tag of the same color, the tags would all respond to the search signal for tag(s) of the color from the interrogator. This situation is clearly not desirable. – When a tag breaks, the user must purchase a replacement tag of the same color as the broken one. Tags are battery-powered. A tag might become a lost object itself after it runs out of battery. More seriously, the interrogator itself can be misplaced. Obviously, these are serious shortcomings.

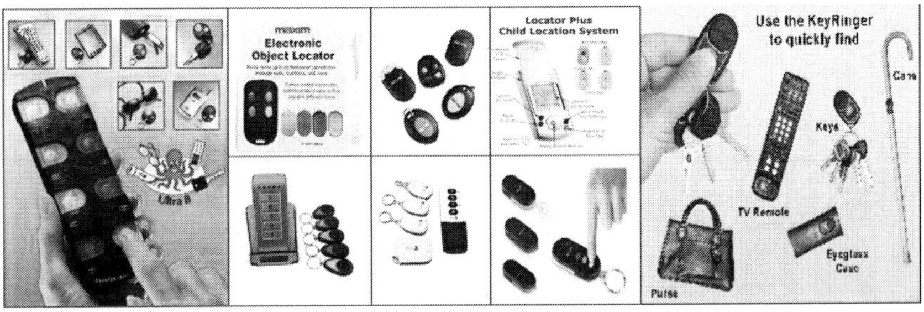

Figure 1. Existing object locators

This chapter describes three designs and a proof-of-concept prototype of object locators based on the*RFID (Radio Frequency Identification)* technology. RFID-based object locators do not have the drawbacks of existing object locators. In particular, RFID-based object locators are extensible, reusable, and low maintenance. They are extensible in the sense that the maximum number of tracked objects is practically unlimited and that a RFID-based object locator can support multiple interrogators. The interrogator software can run on a variety of platforms (e.g. desktop PC, PDA, smart phone and so on). A mobile interrogator can be tagged and thus, can be searched via other interrogators when it is misplaced. Reusability results from the fact that all RFID tags used for object locators can have globally unique ids. Hence, tags never conflict, and a tag can be used in more than one object locators. Low maintenance is one of the advantages of RFID technology. One of the designs uses only RFID tags without batteries; the user is never burdened by the concern that a tag may be out of battery.

This chapter makes two contributions: The first is the object locator designs presented here. The designs use different hardware components and have different hardware-dependent software requirements. The information provided by the chapter on these aspects should enable a developer to build a suitable object locator platform, or an extension to one of the commonly used computer and smart mobile device platforms. The functionality of hardware-independent object locator software is well defined, and a C-like pseudo code description can be found in [9].

The hardware capabilities and object search schemes used by the designs lead to differences in search time and energy consumption. We provide here a numeric model that can be used to determine the tradeoffs between these figures of merit. Developers of RFID-based object locators can use the results of the analysis as design guides. Today, object locators based on all designs are too costly: Typical RFID readers have capabilities not needed by our application and cost far more than what is suitable for the application. Through this analysis, we identify the design that is the most practical for the current state of RFID technology and project the advances in the technology required to make RFID-based object locators affordable (i.e., with prices comparable with some of the locators one can now find in stores.) This is the second contribution of the chapter.

The rest of this chapter is organized as follows. Section 2 describes closely related works. Section 3 describes use scenarios that illustrate how a RFID-based object locator may be used. Section 4 presents three designs of RFID-based object locators. Section 5 describes the implementation of a proof-of-concept prototype based on one of the designs. It also describes the reader collision problem [10] encountered in the prototype and the solution we use to deal with the problem. Section 6 describes a numeric model for computing energy consumption and search time and compares the merits of the designs. Section 7 concludes the chapter and discusses future works.

2. BACKGROUND AND RELATED WORK

This section first presents a brief overview of RFID technology as a way to state the assumptions made in subsequent chapters on state-of-the-art readers and tags. Our object locator resembles location detection systems in its goal: assisting users to locate objects. The section describes existing location systems and compare and contrast them with our object locators.

2.1. Rfid Technology

RFID technology is now applied to a wide spectrum of applications. As an example, personal identification application is used to provide authentication and authorization to individuals carrying their RFID tags so that they can be automatically identified by a central computer. Card-like RFID tags used as smart cards in public transports is another example: Information on money stored in a tag is automatically deducted when the card holder presents the card in front of a reader while getting on or off a transporter. Other applications include using RFID tags as markings of books for more efficient library management, shipping containers for tracking them by retail industry, and so on.

Figure 2 shows a typical system that uses RFID technology. The host machine uses one or more RFID readers to retrieve digital information stored in RFID tags and processes the information according to the needs of one or more applications. In general, a RFID tag contains a globally unique identification (UID) as well as data fields organized in a standard way [11]. A RFID-based object locator only needs the UID information; other data fields are not used.

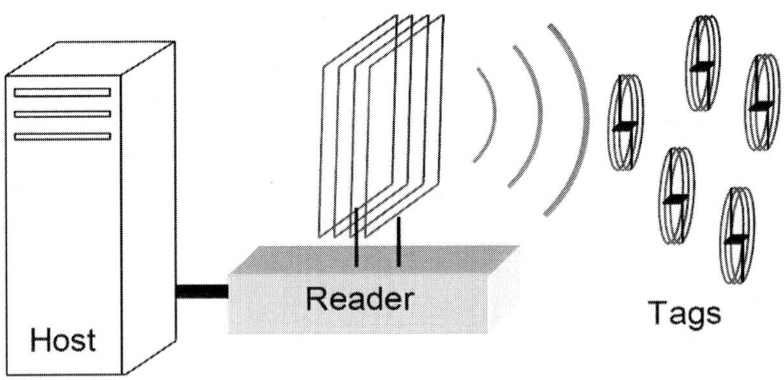

Figure 2. A configuration of RFID system

There are three types of RFID tags: passive, semi-passive and active. A *passive tag* has no internal power source: It gets the power it needs to operate from the incident RF signal radiated by a reader. The readable distance of such a tag ranges from 10 cm to a few meters depending on the frequency of the incident RF signal and its antenna design. In contrast, *semi-passive* and *active tags* have internal power source. Semi-passive tags can increase their readable distances by leveraging internal power. Like passive tags, semi-passive tags respond only after receiving some command from the reader. An active tag, on the other hand, can send RF signals to a reader even when it is not commanded by the reader. Being battery free and having long lifetime (in tens of years) are the major advantages of passive tags over other types of tags for our application.

Each message sent from a reader to tags contains a command code. Among the sets of commands defined by ISO15693 [12], our object locators use only *mandatory* commands and *custom* commands. Standard-compliant tags support all commands in the mandatory set. Commands in the custom set are defined by tag IC manufacturer according to application needs.

The command used to read UID of a tag is the *inventory* command in the mandatory set. This command has only the non-addressed mode, while the other commands have both non-addressed and addressed modes. A command in the *non-addressed mode* is processed by all tags which receive it. A command in the *addressed mode* consists of the command code followed by a UID. When a tag receives an addressed-mode command, it first checks whether the UID is its own. The tag processes and responds to the command only when it is the tag addressed by the UID.

2.2. Location Detection Systems

Many different location detection systems are available today. Global Positioning System (GPS) [13] is the most well known. Priced at about $ 100 US each, GPS navigators are widely used in cars, buses and so on. However GPS has its limitations. Reflection, occlusion and multipath effects seriously interfere with distance measurement and make GPS ineffective indoors. For this reason, indoor location detection systems use a variety of other technologies.

Active Badge [14] is representative of infrared-based location detection systems. A badge containing an infrared transmitter is attached to each object to be tracked by the system. The transmitter sends periodically messages containing the unique identification of the badge. The messages are caught by some infrared receivers at fixed known locations and relayed by the receivers to a central computer. The central computer resolves the position of the badge based on the locations of the receivers. Shortcomings of systems such as Active Badge arise from the fact that infrared signals cannot penetrate most materials in a building and are easily interfered by other infrared sources.

Ultrasound is used to assist with distance measurement in Bat [15] and Cricket [16]. These systems use both ultrasound and RF signals to measure distances between beacons (transmitters) and listeners (receivers): When a beacon at a known location transmits an ultrasound signal and a RF signal concurrently, a listener can calculate the distance to the beacon from the difference between the arrival times of the signals.

Many indoor location detection systems use RF-based technology to take advantage of the fact that RF signals penetrate most non-metallic materials. RADAR [17] is an example. The system estimates distance by estimating the strength of RF signals. Specifically, the system measures in the initialization phase at a set of fixed locations the strengths of a RF signal sent by a location-known transmitter. The measured strengths are stored in a database to be used later as yardsticks during the working phase. In the working phase, each receiver measures the strength of a RF signal transmitted from a tracked object and sends the strength to a central computer. The computer compares the measured strengths with the information stored in the database and then resolves the possible position of the transmitter (i.e., the tracked object). MoteTrack [18], similar to RADAR, uses empirical distance measurement to estimate positions of objects. WLAN (wireless local area network) can be used to build location detection systems also. SpotON [19] and Nibble [20] are examples.

Compared with the above mentioned location detection systems, an object locator must be a far more low cost solution and must be ultra easy to set up and use. Many indoor location detection systems (e.g. Bat and Active Badge) rely on a big infrastructure or a pre-computed database (e.g. RADAR) to support location estimation. These systems are too costly to deploy and maintain and hence, unsuitable for home use. Cricket system provides a low cost location-aware service. An object with a receiver can determine its location. This is not what an object locator does. A misplaced object does not need to know its own location; the user looking for it needs to know.

3. USER SCENARIOS

The routine usage of an object locator requires only three operations: Add, Delete and Query. We describe these operations here to illustrate how a locator may be used. Without loss of generality, we assume that a new object locator kit contains a portable interrogator, a dozen of RFID tags and agents. As illustrated by Figure 3, the interrogator resembles a smart phone. It has a small non-volatile storage and a RF transceiver together with a network address. We will return in the next section to describe how the RF transceiver is used, as well as what agents are and do. Unlike common smart phones, however, the interrogator has a RFID reader. The reader is used for the Add operation described below.

Specifically, Figure 3 shows parts of the user interface on an interrogator with a LCD touch screen and two buttons. The LCD touch screen is used as both input and output user interface. A user can select an item among the items displayed on the screen, the button at the bottom left corner to confirm a selection, and the button at the bottom right corner to cancel the selection. Some operations need text input. The virtual keyboard shown on right is for this purpose.

Figures 4 and 5 illustrate Add and Query operations, respectively. Add operation works in a similar way as the address book of a smart phone. Using this operation, the user can add the registration of an object to be tracked into the interrogator. By registration, we mean a mapping between the id of the tag attached to an object and the name of the object. The user queries the locations of objects by their names. In response to a query, the interrogator uses the object-name-tag-id mappings to resolve which one of the registered objects to search. Figure 4 shows a scenario: The user picks an unused tag and attaches it to an object to be tracked as shown in Figure 4(a) and (b). Then, the user puts the tag close to the interrogator and selects Add object. This step is shown in Figure 4(c). In response to Add object command, the interrogator reads the id of the tag, displays a new text field and prompts the user to enter a name (e.g., Key). When the user confirms the name, the interrogator creates a mapping associating the name with the id of the tag attached to the object, and stores the mapping in its local non-volatile memory. This is illustrated in Figure 4(d). The user repeats the above steps to register each object until all objects to be tracked are registered.

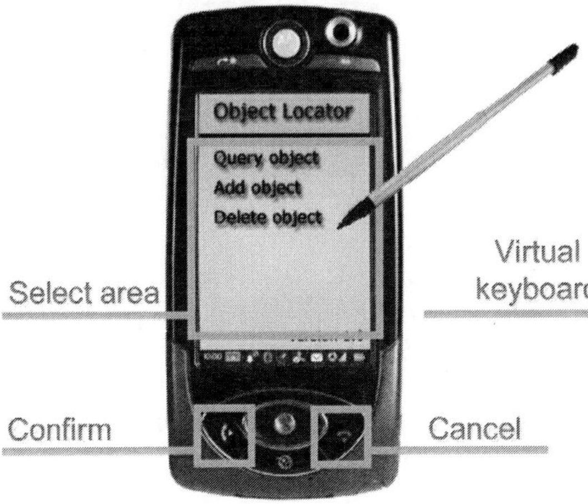

Figure 3. Object locator user interface

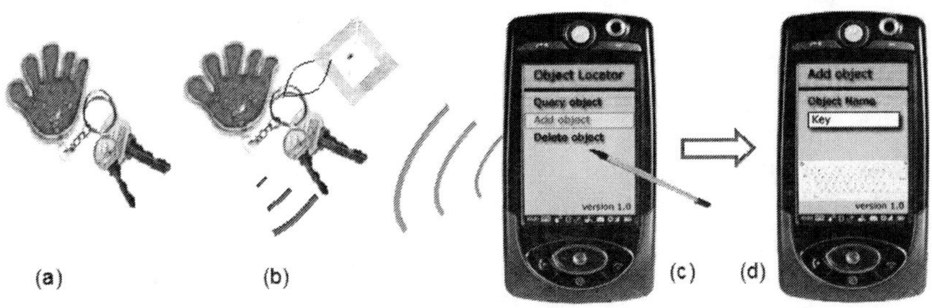

Figure 4. Add operation

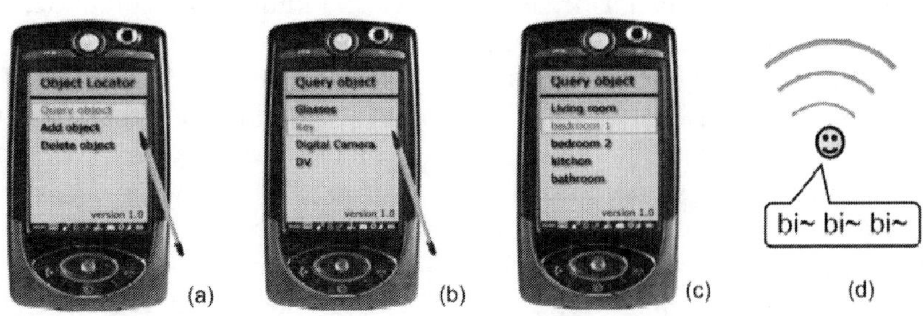

Figure 5. Query operation

Query operation is the work horse of the object locator. The user presses Query object on the touch screen, as illustrated by Figure 5(a), to invoke this operation for assistance in finding misplaced objects. When the names of registered objects are displayed, the user selects the object to be searched; in this example, it is Key. After the user confirms the selection, as shown in Figure 5(b), the interrogator retrieves from its local storage the id of the tag attached to the object with the selected name and starts a search for the tag with that id. Hereafter, we call the tag being searched the *queried tag* and the object attached to the tag the *queried object*. We will describe the search process in the next section.

Object locators of different designs present the result of Query operation in different ways. As examples, Figure 5(c) and (d) shows two different responses. In Figure 5(c), the interrogator directs the user to the place (e.g. bedroom 1) where the queried object is found. In Figure 5(d), the queried tag beeps, allowing the user to look for it by following the sound. This version works like the existing locator described in Section I.

Delete operation removes the registration of an object, i.e., the object-name-tag-id mapping stored in the interrogator: The user can invoke the operation by pressing Delete object on the touch screen. In response, the interrogator displays the list of registered objects, allowing the user to select the object (e.g. Key) to be deleted. The interrogator deletes the mapping after the user confirms the selection. Delete operation frees the tag attached to the now unregistered object and makes the tag free for use to track some other object.

4. ALTERNATIVE DESIGNS

The three designs of object locator are called Room-level Agents, Interrogator and Tags (RAIT) locator, Desk-level Agents, Interrogator and Tags (DAIT) locator and Desk-level and Room-level Agents, Interrogator and Tags (DRAIT) locator. As their names imply, each of the locator consists of tags, agents and at least one interrogator. The adjectives room-level and desk-level describe the ranges of RFID readers used by the designs. The ranges of room-level readers and desk-level readers are sufficiently large to cover a typical-size room or desk, respectively.

The term tag refers specifically to RFID tags. Each tag has a unique id, hereafter called *TID*. One of the designs uses only passive tags. The other designs call for tags that can beep upon receiving query messages containing their TIDs. It is possible to implement such tags using semi-passive RFID tags since the battery in such a tag can be used not only to improve read range but also to drive a beeper.

An *agent* is a device that aids the interrogator in locating the queried object (i.e., the queried tag). Each agent has a RF transceiver, together with a programmable network address, a RFID reader, and a RFID tag. The RFID reader in the agent enables the agent to search for the tags within its coverage area. As stated in Section III, the interrogator also has a RF transceiver with a network address. This allows the interrogator and all agents to form a wireless local area network (WLAN). The network address of the interrogator (or each interrogator in a multiple-interrogator system) is unique and so is the network address of each agent. The interrogator requests assistance from an agent by sending the TID of the queried tag to the agent via the WLAN. We assume that the network provides reliable communication. We do not mention other aspects of the WLAN because they are not relevant to our discussion.

4.1. Rait Locator

A disadvantage of the existing locator is that a user needs to walk around the house when searching an object and the interrogator needs to repeatedly send the query signal until the user hears the queried tag or gives up the search. RAIT locator is designed to eliminate this disadvantage.

RAIT locator uses one or more agents to cover each room, and the house is fully covered by agents as shown in Figure 6. When the user invokes a Query operation, the interrogator sends a query message containing the TID of the queried tag to agents and thus requests the agents to search the queried tag on its behalf. Each agent broadcasts an addressed mode read request with the TID retrieved from the query message to read the tags within range. The tag with id matching the TID beeps upon receiving a read request, in addition to responding to the agent. The agent finding the queried tag reports its network address to the interrogator. This information enables the interrogator to display the results illustrated by Figure 5(c), telling the user to go to the specified room where the queried object has been found.

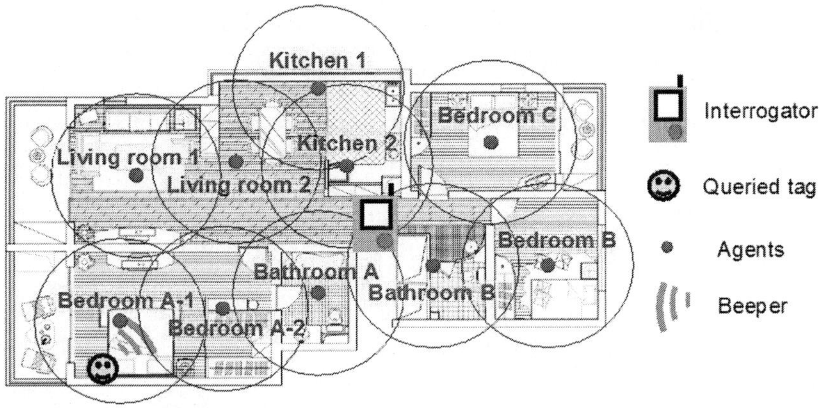

Figure 6. Configuration of RAIT locator

Obviously, the agents must be set up before a RAIT locator can be used. Figure 7 lists the steps carried out by the user and work done by the system during the set up process. The goal of Steps 3-5 is to make sure that there is no blind region. A *blind region* is an area where tags cannot be read by any agent. The corners of a room are the most likely to be blind regions. This is the rationale behind Step 3. When the *TEST READ RANGE* switch of an agent is on, the agent repeatedly broadcasts non-address mode read messages. In this way, the agent enables the user to determine whether any of the corners is a blind region in Step 3.

1. Choose a location near middle of a room and temporarily attach an agent to the ceiling or furniture at the location.
2. Turn on *TEST READ RANGE* switch on the agent.
3. Pick up a tag and check whether the tag beeps at each corner of the room.
4. If no, adjust the location of the agent or add one more agent at another location in the room and turn on *TEST READ RANGE* switch on the additional agent. Then go back to Step 3. If yes, turn off *TEST READ RANGE* switch.
5. Securely attach the agents tested in Steps 2-4 at their respective locations.
6. Put the interrogator near the agent and execute *Register Agent operation*.
7. Repeat step 1 to 6 until all agents covering the house are registered.

Figure 7. Agent set-up process

The Register Agent operation in Step 6 is similar to Add operation described in Section 3. Its goal is to assign a human-readable location name to an agent, so that the interrogator can later generate query results illustrated by the example in Figure 5(c). During the operation, the interrogator prompts the user to provide a unique name for the location of each agent. For example, if the living room needs two agents, Living Room R(ight) and Living Room L(eft) are good names for them.

The interrogator also assigns a unique network address to the agent being registered. The id of the tag in an agent is the product serial number of the agent. The interrogator uses the id to distinguish the agent from previously registered agents. By assigning successive network addresses to agents as they are registered and initialized one by one, successive Register Agent operations enable each initialized agent to join the WLAN and later compute the addresses of other agents by adding or substituting some number from its own address.

Figure 8 depicts the format of messages in a RAIT locator. This format supports multiple interrogators: The src_addr allows agents to identify the interrogator issuing the query message. The dest_addr allows them to address their responses to a specified interrogator. Data field allows interrogators to synchronize their databases created by Add and Register Agent operations. We will discuss how the other fields are used shortly.

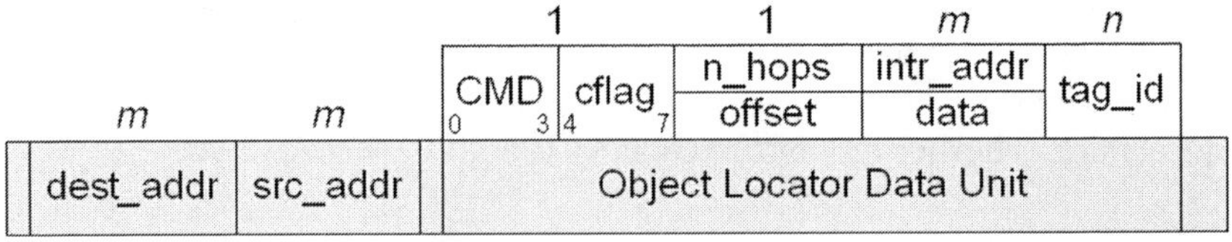

Figure 8. RAIT locator message format

4.2. Dait and Drait Locators

DAIT locator, shown in Figure 9, is an extension of RAIT locator. The designs are similar in how the Query operation is handled by the interrogator and agents. DAIT differs from RAIT primarily in the required read ranges of agents. The read range of agents used in a DAIT locator is less than one meter. Agents with such a small range offer higher accuracy in locations of queried tags. Information on the agent that finds the queried tag tells the user the location of the searched object within a small vicinity of the agent. Tags in DAIT locators are passive; they do not beep because a user can easily find the misplaced object even though the tag does not beep. Because tags do not need to beep, they can be battery free. This is a major advantage of DAIT locator.

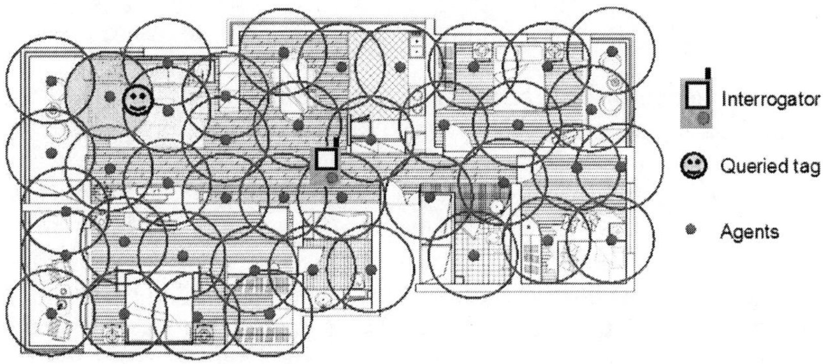

Figure 9. Configuration of DAIT locator

However, it is significantly more complicated to set up desk-level agents. Blind regions of RAIT locator are easy to detect and eliminate because a blind region is typically created by walls and is near the read boundary of an agent. In the case of DAIT locator, a room cannot be fully covered by one or two agents. Any three adjacent agents may create a blind region. Our solution is to give a user a circular thread whose circumference is less than $\sqrt{33}$ (i.e., the circumference of a regular triangle whose center is one unit away from its corners) times their read range and instruct a user to set any three adjacent agents within the circular thread. By doing so, blind regions never occur.

DRAIT locator has a hybrid design that aims to extend the lifetime of semi-passive tags. A DRAIT locator contains both room-level and desk-level agents. Its interrogator asks desk-level agents to search first. The interrogator asks room-level agents only when no desk-level agent finds the queried object. We set up desk-level agents on furniture in addition to setting up room-level agents as described above. Because misplaced objects are often on furniture or in vicinities of them, the queried object can often be found by a desk-level agent, and the tag on it does not need to beep.

4.3. Search Schemes

A queried object can be searched in three ways: broadcast, relay and polling. The *broadcast scheme* is the most straightforward. The interrogator broadcasts a query message with the tag_id field filled with TID of the queried tag. The agents finding the queried tag report their agent ids to the interrogator and the others do not reply.

The knowledge on the agent network addresses and the number of agents enables an interrogator to request assistance from agents one at a time using the *relay scheme*: To search for a queried tag, the interrogator sends a query message containing its own address in intr_addr field, the number of agents to be queried in n_hops and the TID of the queried tag in tag_id to the first agent: The simplest choice is the agent with the smallest address. In response to a query, each agent searches for the tag with the TID in its own cover area. The agent reports its own address to the interrogator if it finds the tag; otherwise it decreases n_hops by one, increments its own network address by one to get the address of the next agent and then forwards the query message to the next agent.

According to the *polling scheme*, the interrogator also sends a query message to the first agent in its polling list, provides the agent with the TID of the queried tag and waits for response from the agent. The agent replies to the interrogator no matter whether it finds the tag or not. If the response from an agent is negative, the interrogator sends the query message to the next agent in its polling list. Advantage of the polling scheme over the relay scheme is that the interrogator can dynamically alter the search sequence.

5. PROTOTYPE IMPLEMENTATION

We implemented a proof-of-concept prototype of DAIT locator, the design that does not require customized semi-passive tags. Indeed, all components used in our prototype are readily available today. Parts (a) and (b) of Figure 10 show an agent and the portable interrogator of our prototype, respectively. The agent is composed of a microcontroller, a RF transmitter, a RF receiver and a RFID reader module. The microcontroller is ATMEL ATmega128. It runs at 8MHz and has 128k bytes flash / 4k bytes EPPROM. The RF transmitters and receivers interconnecting interrogator(s) and agents are LINX TXM(RXM)-433-

LR, which use 433MHz ASK. RFID reader modules are MELEXIS EVB90121, which is ISO15693-compliant and uses a directional antenna. We use TI OMAP5912 and NEC Q-VGA to implement the portable interrogator. The current version of our prototype supports the three operations described in Section II and uses the polling search scheme.

The lack of customized antenna design for tags and readers and the reader collision problem seriously affects the performance of our prototype. Our DAIT prototype uses only tags with directional antennae. (Again, the reason is that such tags are readily available.) When the antennae of tags and readers are directional, the read performance of agents depends on the orientation of the antennae. Clearly, tagged objects may be placed in arbitrary orientations. As a consequence, it is impossible to ensure optimal or near optimal alignment of the tag antennae towards the agents covering their locations. This is the reason that tags in a DAIT object locator should have omni-directional antennae. Agents with omni-directional antennae can be simply set on furniture as shown in Figure 11(a). Agents with directional antennae should be attached to the ceiling as shown in Figure 11(b). This arrangement requires a read range of 2-3 meters. With readers of a sufficiently large read range, RAIT locators can use tags with directional antennae without performance concern.

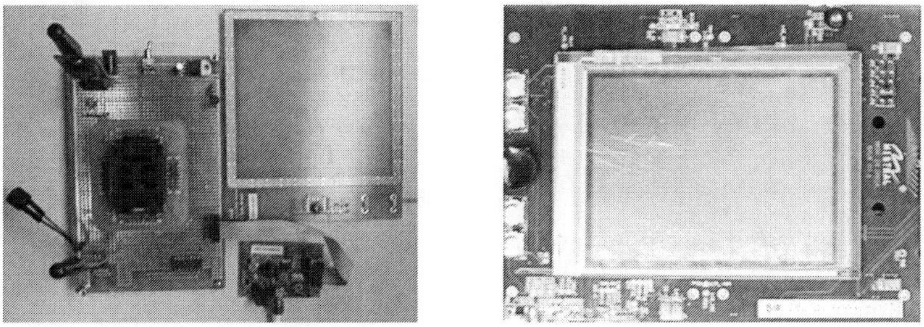

(a) (b)

Figure 10. Agent and interrogator

Close proximity of readers (i.e., agents) is necessary in order to avoid blind regions. Our DAIT prototype is no exception. When RFID readers have overlap coverage areas, signals sent at the same time from them to tags in the overlap region interfere with each other. This is called the *reader collision problem* [10]. Fortunately, only the broadcast scheme suffers this problem. Our prototype uses the polling scheme to avoid the problem: According to the polling schemes (or the relay scheme), agents search the queried tag in sequence; signals from readers never interfere.

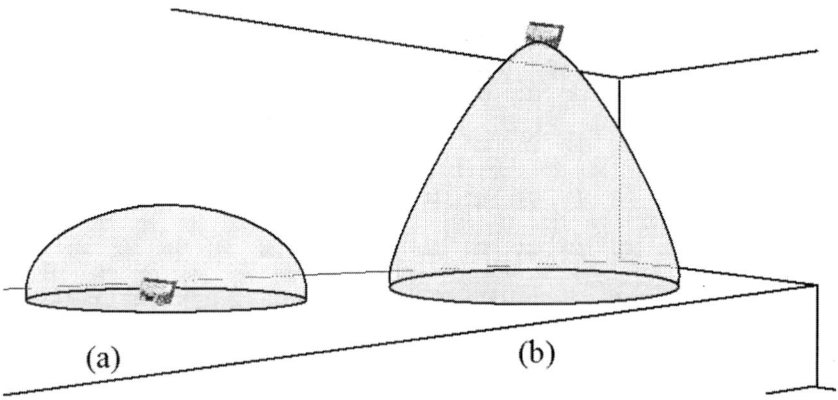

Figure 11. Arrangement of agents

A DAIT locator that uses the broadcast scheme can circumvent the reader collision problem in many ways. For example, the DAIT prototype can let each agent delay transmitting its query signal by an amount of time that is a function of its network address. In this way, agents try to avoid transmitting query signals at the same time. This solution is practical and easy to implement.

Another solution requires each agent to know the network addresses of its neighbors. Each agent can be viewed as a node in a connected graph. There is an edge between two nodes when the agents represented by them have overlapping coverage regions. A graph coloring algorithm can be used to assign different colors to adjacent nodes. The reader collision problem never occurs as long as agents labeled by different colors do not transmit query signals concurrently. This solution is likely to have a better response time than the solution mentioned above or the relay and polling schemes. However it requires additional hardware for each agent to automatically detect its neighbors or connectivity information entered by the user manually. The additional hardware makes agents more costly, and complicated operations by the user make an object locator hard to use.

6. RELATIVE MERITS

We use search time and energy consumption of a single query to measure the relative merits of object locator designs. Search time and energy consumption per query depend on many factors including the number of agents, search scheme, search sequence and locations of misplaced objects.

6.1. Search Time and Energy Consumption

The expressions of energy consumption and search time per query according to broadcast, relay and polling schemes are listed in Table 1. The expressions assume that agents and interrogator(s) are battery powered and communicate in the manners described in Section 4. The notations used in the expression are defined in Table 2.

The total energy consumed by the object locator for processing a Query operation according to the broadcast scheme is the sum of the three terms in the first row of Table 1. In this case, the interrogator transmits only one query message per Query operation. The energy it consumes is *EIA*. The energy consumed by each agent in the search is *EArfid*. The total energy consumed by all agents is *NA*(*x, y, r*)*EArfid*, where *NA*(*x, y, r*) is the number of agents with range *r* in a rectangular space of dimensions *x*and *y*. The agent finding the queried tag consumes *EAI* to send a response back to the interrogator.

In the expressions, *pAi* denotes the probability that the *i*-th agent in the search sequence finds the queried tag. In general, this probability is a function of the number and location distribution of objects (i.e., tags) in the house. (To keep the expressions simple, our notations do not show this dependency.)

The expression of the expected time taken by the locator using the broadcast scheme to respond to a Query operation assumes that agents search the queried tag in sequence in order to avoid the reader collision problem. The first term in the expression is the time taken by the query message from the interrogator to reach all the agents. If the first agent finds the queried tag, which occurs with probability *pA1*, the addition delay is *DArfid* + *DAI*. This is the reason for the second term in the expression of *Tavg*. In general, the probability that the queried tag is found by the *i*-th agent is $\prod_{k=1}^{i-1}(1-pA_k)pA_i$. When this occurs, each of the other agents spends $DArfid$ amount of time to search for the queried tag before the i-th agent can respond to the interrogator. Hence, the delay is $iDArfid + DAI$.

Table 1. Expressions for search time and energy consumption

broadcast	E_{total}	$E_{IA} + N_A(x, y, r)E_{Arfid} + E_{AI}$
	T_{avg}	$D_{IA} + pA_1(D_{Arfid} + D_{AI}) + \sum_{i=2}^{n}(\prod_{k=1}^{i-1} 1 - pA_k)pA_i(iD_{Arfid} + D_{AI})$
relay	E_{avg}	$E_{IA} + pA_1(E_{Arfid} + E_{AI}) + \sum_{i=2}^{n}(\prod_{k=1}^{i-1} 1 - pA_k)pA_i(iE_{Arfid} + (i-1)E_{AA} + E_{AI})$
	T_{avg}	$D_{IA} + pA_1(D_{Arfid} + D_{AI}) + \sum_{i=2}^{n}(\prod_{k=1}^{i-1} 1 - pA_k)pA_i(iD_{Arfid} + (i-1)D_{AA} + D_{AI})$
polling	E_{avg}	$E_{IA} + pA_1(E_{Arfid} + E_{AI}) + \sum_{i=2}^{n}(\prod_{k=1}^{i-1} 1 - pA_k)pA_i(iE_{Arfid} + (i-1)E_{IA} + iE_{AI})$
	T_{avg}	$D_{IA} + pA_1(D_{Arfid} + D_{AI}) + \sum_{i=2}^{n}(\prod_{k=1}^{i-1} 1 - pA_k)pA_i(iD_{Arfid} + (i-1)D_{IA} + iD_{AI})$

Table 2. Notations

- D_{IA}: Delay of a message transmitted from an interrogator to an agent
- E_{IA}: Energy consumption of a message transmission from an interrogator to an agent
- D_{AA}: Delay of a message transmitted from one agent to another
- E_{AA}: Energy consumption of a message transmission from one agent to another
- D_{AI}: Delay of a message transmitted from an agent to an interrogator
- E_{AI}: Energy consumption of a message transmission from an agent to an interrogator
- D_{Arfid}: Time for an agent to use its RFID reader to search a queried tag
- E_{Arfid}: Energy consumption of a RFID reader in an agent per search of a queried tag

The average search time of an object locator that uses the relay and polling scheme are estimated by the expressions in the fourth and sixth rows in Table 1, respectively. Relay and polling scheme also lets all agents search the queried tag in sequence. This is why the coefficients in these expressions are the same as the coefficients in the expression of *Tavg* for the broadcast scheme. The expressions of the average energy consumption can be derived from the expressions of the average search time by substituting energy consumption for message transmission delay because sending a message cause both transmission delay and energy consumption.

As stated earlier, Table 1 is based on the assumption that agents and the interrogator are battery powered. Hence, the total energy consumption includes energy consumptions of agents and an interrogator. However, agents can be connected to wall plugs, especially when the number of agents is small, as in the case of RAIT locators. The interrogator using relay and broadcast scheme consumes exactly *EIA* to search a queried tag. The interrogator using polling scheme consumes at least *EIA* to search a queried tag. Therefore, the polling scheme is suitable for stationary interrogator(s) and the relay and broadcast scheme are suitable for portable interrogator(s) if we do not need to account for the energy consumption of agents.

6.2. Model Of Object Locality

The probability *pAi* of that an agent *Ai* finds the queried tag, and hence the misplaced object, depends on where the object is at the time. To calculate this probability, we use a locality model of tracked objects. The model gives the spatial probability density of the locations of each object. For the sake of simplicity and without noticeable lose of accuracy, we partitions the space in the search area into unit squares, rather than treating the coordinates of a location as continuous variables. (Except for where it is stated otherwise, the dimension of a unit square is 1 cm by 1 cm.) This allows us to model a house as a finite, discrete and planar search space. We denote the space by Z={Zx,y}⊆N×N. Each element *Zx,y* of the space is a unit square; its location is given by the coordinate (x, y) where both *x*and *y* are integers. All agents are at fixed and known locations. A misplaced object may be placed anywhere within the search space.

We call the probability of finding a queried object at *Zx,y* the (*existence*) *probability* of the object at*Zx,y*. (For example, if we find an object at *Zx,y* on the average 10 times in 100 searches for the object, the (existence) probability of the object at *Zx,y* is approximately 0.10. We use *pZx,y*(*j*) to denote the existence probability of an object with a tag of id = *j* at *Zx,y*. We do not consider the situation where someone has taken some registered object shopping, for example, while someone else is searching for it in the house. Hence, for every object being searched, the sum of the probabilities of it being at all locations in the search space equals to 1.

Figure 12 gives an illustrative example. The figure is not drawn scale, and each unit square in this example is 10 cm by 10 cm in dimension. Two agents *A1* and *A2* are at their locations. The id of *A1* is 1 and the id of *A2* is 2. The rectangle models a desk. It contains 15 unit squares. The number in each square gives the probability of a queried object being at the location. Since the numbers add up to 1, they tell us that the object is surely somewhere in the rectangle. We want to calculate *pAi*, the probability that the agent with id = *i* can find the queried tag. Using Figure 12 as an illustrative example, we see that *pA1* equals to the sum of all existence probabilities within the read range of the agent *A1*; in other words, *pA1* is about 0.87. Similarly, we find that *pA2* is about 0.68.

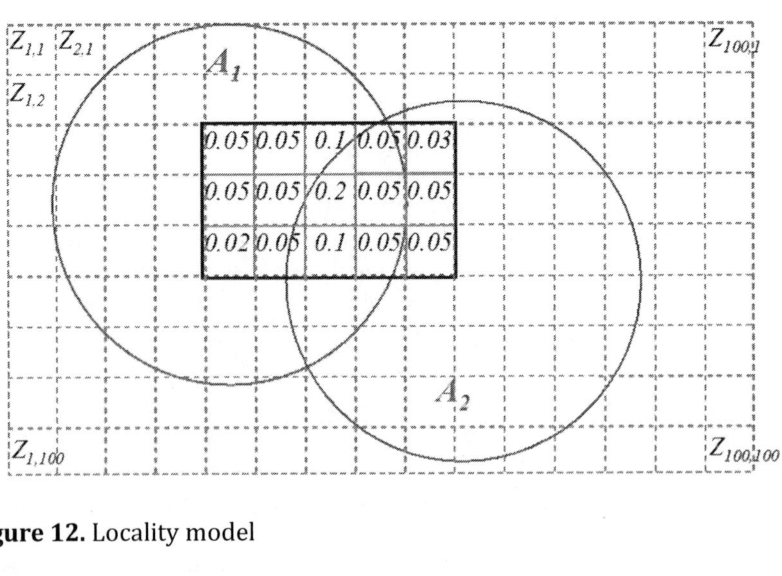

Figure 12. Locality model

We call the area where a misplaced object might be placed an *object region*. The size of an object region is the total area of the region in number of unit squares. We characterize the locality of a misplaced object by the size and shape of its object region and its existence probabilities of being at each unit square within the region. Once we know the locality parameters of an object and coverage area of each agent *Ai*, the terms *pAi* can easily be calculated. We can then calculate the average search time and energy consumption of the object based on the probability *pAi* for all agents.

6.3. Evaluation Environment and Results

The environment we used to evaluate the relative performance of our designs has a 10m by 10m search space, containing 1000 × 1000 unit squares of size 1 cm by 1 cm. Agents are placed according to the arrangement in Figure 13(a). The number of agents is *NA(1000, 1000, r)*. Again, *r* is the read range of an agent. The ranges of desk-level and room-level agents are 100 and 350, respectively, the typical number of room-level agents in a RAIT locator is *NA(1000,1000,350)* = 6, and the typical number of agents in a DAIT locator is equal to *NA(1000, 1000,100)* = 42.

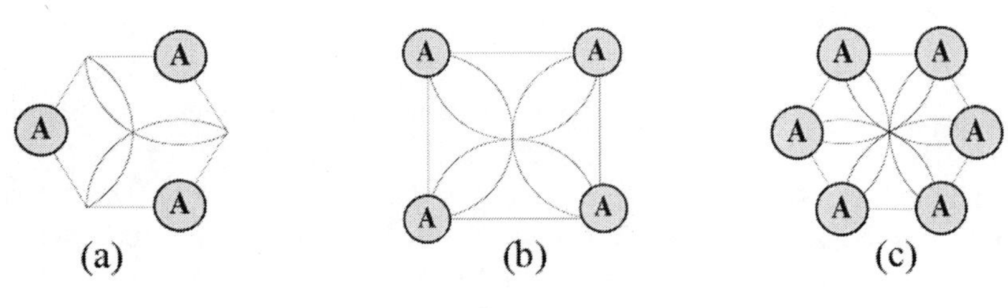

Figure 13. Possible arrangement of agents

In Section 4, we said that the agent with the smallest network address is the first agent and the other agents are asked one by one in order of agent ids to search for the queried object. We call this search order *sequential*. Alternatively, we can ask the agents in non-increasing order of their empirical existence probabilities. This search sequence is called *profiling*.

Our evaluation program assumes that object regions are circular for the sake of simplicity. The center and radius of an object region are randomly generated. The variables *DIA, DAA* and *DAI* in Table 2have the same values because both interrogators and agents use the same kind of RF transceiver. For the same reason, *EIA*, *EAA* and *EAI* have the same value. For convenience, we use *DArfid* and *EArfid* as base units of delay and energy consumption. The ratio of *DIA/DArfid* (*DAI/DArfid* and *DAA/DArfid*) is called *DRatio* and the ratio of *EIA/EArfid* is called *ERatio*. The evaluation program needs only these two parameters rather than all variables.

Figure 14(a) and (b) show the average search time for broadcast scheme, relay scheme, and polling scheme (i.e., polling in sequential order), as well as polling scheme with profiling. The search time of relay and polling schemes is higher than broadcast scheme for all values of DRatio. The search time of polling scheme with profiling is less than that of broadcast scheme when DRatio is less than about 1 (10^{0}) for *NA* = 42 and 1.25 ($10^{0.1}$) for *NA* = 6.

Figure 15 shows the average energy consumption consumed by agents when *NA* is 42 and 6. The energy consumption consumed by agents is the same, when the relay and polling scheme is used. AsFigure 15(a) depicts, the energy consumption of relay scheme and polling scheme are the same. Their consumptions and that of polling scheme with profiling is less than that of broadcast scheme when ERatio is less than 1.99 ($10^{0.3}$) and 7.94 ($10^{0.9}$), respectively. Values of ERatio at the intersections of the curves in Figure 15(b) are about 3.16 ($10^{0.5}$) and 15.85($10^{1.2}$).

Table 3 gives a summary. The table suggests the broadcast scheme when DRatio is high and search time is more important than energy consumption. When DRatio is low, the differences among the search times of all search schemes are small. Energy consumption becomes the dominant factor for comparison. It is possible for agents in a RAIT locator to connect to power source. For energy saving on interrogators, we suggest polling scheme with profiling for stationary interrogators and relay or broadcast scheme for portable interrogators. As for DAIT locators, we consider energy consumption of an interrogator and agents. We suggest polling scheme with profiling when ERatio is low and the same suggestions as that for a RAIT locator if ERatio is high.

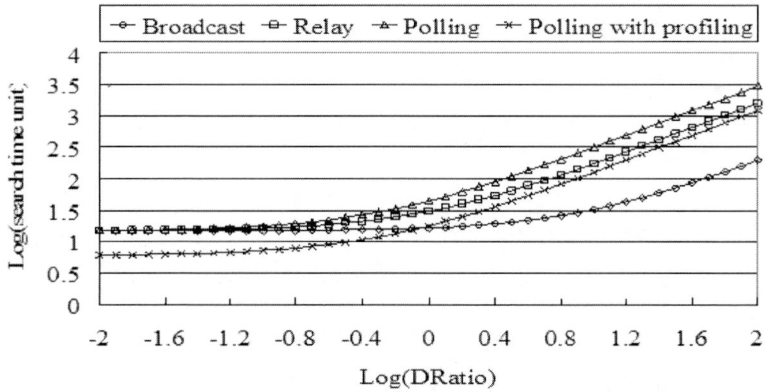

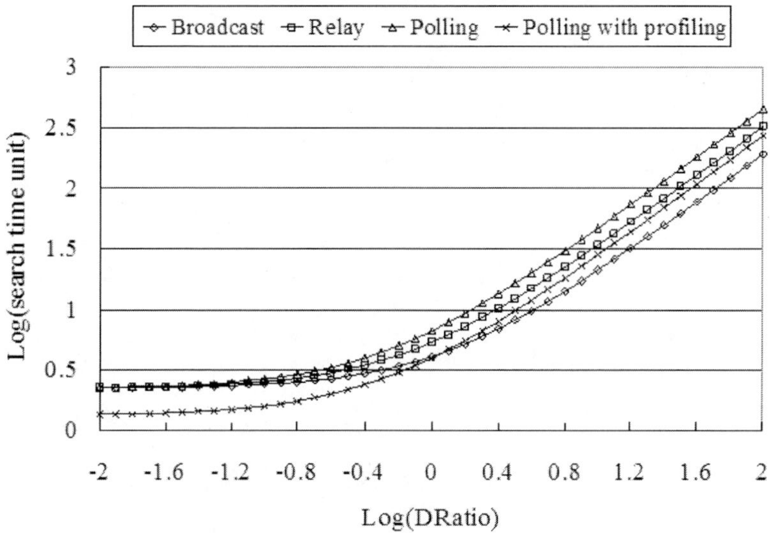

Figure 14. Search time Vs DRatio: (a) top NA = 42; (b) bottom NA = 6

7. CONCLUSION

We described here three alternative designs for RFID-based object locator. These object locators are extensible, reusable and low maintenance. They are easy for users to set up and use. Our analysis shows that search time and energy consumption for all designs and search schemes depend the capabilities of RFID readers and RF transceivers used by agents. Roughly speaking, polling and relay schemes are competitive to broadcast scheme only when DRatio or ERatio are less than 10.

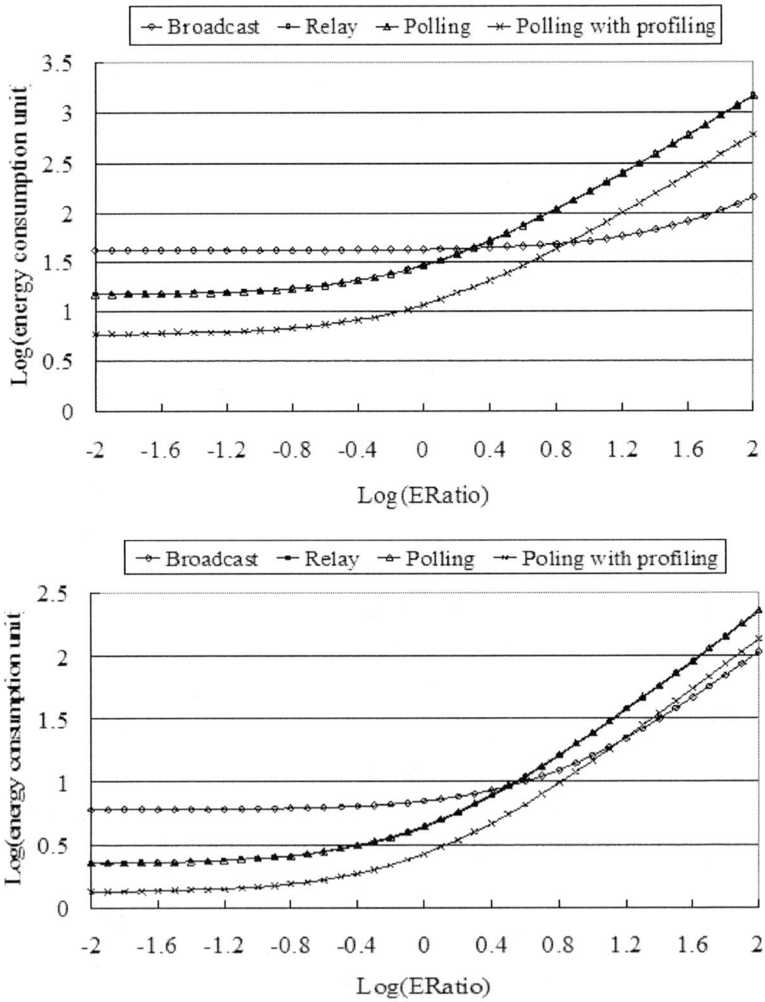

Figure 15. Energy consumption of agents Vs ERatio: (a) top NA = 42; (b) bottom NA = 6

Table 3. Summary of suggested search schemes

	Low DRatio Low ERatio	Low DRatio High ERatio	High DRatio Low ERatio	High DRatio High ERatio
RAIT locator (fewer agents)	PI: ***broadcast*** or ***relay*** SI: ***polling with profiling***	PI: ***broadcast*** or ***relay*** SI: ***polling with profiling***	***broadcast***	***broadcast***
DAIT locator (more agents)	***polling with profiling***	PI: ***broadcast*** or ***relay*** SI: ***polling with profiling***	***broadcast***	***broadcast***

PI: portable interrogator SI: stationary interrogator

PI: portable interrogator; SI: stationary interrogator

We implemented a proof-of-concept DAIT prototype object locator to demonstrate the object locator concept and designs. The prototype uses only readily available hardware components, including readers and tags with directional antennae. The performance of the prototype is far from ideal, primarily for this reason. Because it is impossible to control the orientation of tag antennae, omni-directional antennae are better suited for our application.

The total cost of an object locator depends on many factors. The total hardware cost of a minimum object locator is the sum of the costs of an interrogator and required number of agents and tags. Compared with the costs of interrogator and agent, the hardware cost of tags is significantly lower and, for the discussion here, can be neglected.

Currently, the total hardware cost of an object locator is dominated by the total cost of agents, and the cost of an agent is dominated by the RFID reader in the agent. The number of agents required to fully cover a house depends on dimensions *x* and *y* of the house, the read range of the agents and the way agents are placed. To get a rough estimate, we assume that the coverage area of each agent is a circle.Figure 13 depicts three ways to place agents. Putting agents further apart than locations shown inFigure 13(a) can create blind regions. Putting more agents closer than those indicated in Figure 13(c)is not necessary since the space is covered by at least two agents. We need six room-level agents to cover a 10m x 10m space even when we place agents as shown in Figure 13(a) (i.e., as far as possible without creating blind regions). The existing object locator costs $ 50 US. A RAIT locator is not competitive to the existing locator unless the cost per room-level agent is about $ 10 US. As for DAIT locator, the cost per desk-level agent must be much lower. We are optimistic that the cost of agents will become sufficiently lower in the coming decade as the need for more and more products (e.g., Smart pantry [1], dispenser in [4]) containing RFID readers are developed to take advantage of this technology.

ACKNOWLEDGMENT

This work is partially supported by the Taiwan Academia Sinica thematic project SISARL (Sensor Information System for Active Retirees and Assisted Living (http://www.sisarl.org).

REFERENCES

1. Hsu CF., Liao HY., Hsiu PC., Lin YS., Shih CS., Kuo TW., Liu JWS., Smart Pantries for Homes: Proceedings of IEEE International Conference on Systems, Man and Cybernetics, SMC2006, 8-11 October 2006, Taipei, Taiwan; 2006. p4276-4283.

2. Yeh HC., Hsiu PC., Tsai PH., Shih CS. and Liu JWS. APAMAT: A Prescription Algebra for Medication Authoring Tool: Proceedings of IEEE International Conference on Systems, Man and Cybernetics, SMC2006, 8-11 October 2006, Taipei, Taiwan; 2006. p3676-3681.

3. Hsu YT., Hsiao SF., Chiang CE, Chien YH., Tseng HW., Pang AC., Kuo TW., Chiang KH. Walker's Buddy: An Ultrasonic Dangerous Terrain Detection System: Proceedings of IEEE International Conference on Systems, Man and Cybernetics, SMC2006, 8-11 October 2006, Taipei, Taiwan; 2006. p4292 – 4296.

4. Tsai PH., Yu CY., Shih CS., Liu JWS. Smart Medication Dispensers: Design, Architecture and Implementation. IEEE Systems Journal 2011; 5(1). p99-110.

5. Medication Reminders: http://www.epill.com/ (accessed 01 July 2012)

6. Senior Care Products: http://www.agingcare.com/Products (accessed 01 July 2012)

7. Alababa.com, showroom for object locators: http:// www. alibaba.com/showroom/object-locator.html (accessed 01 July 2012)

8. Hsu CC., Chen JH. A novel sensor assisted and RFID-based indoor tracking system for elderly living alone: Sensor 2011; 11. p10094-10113.

9. Chou TS. Design and Implementation of Object Locator. MS thesis, Taiwan National Tsing Hua University, 2006

10. Leong Kin Seong LK., Ng ML., Cole PH. The Reader Collision Problem in RFID Systems: Proceedings of IEEE International Symposium on Microwave, Antenna, Propagation and EMC Technologies for Wireless Communications, MAPE 2005, August 2005. p 658 – 661.

11. EPCglobal standards overview: http://www.gs1.org/ gsmp/kc/ epcglobal (accessed 11 July 2012)

12. ISO15693 Standard, OpenPCD: http://www.openpcd.org/ISO15693 (accessed 12 July 2012)

13. Getting I. The Global Positioning System: IEEE Spectrum 1993; 30(12). p36–47.

14. Want R., Hopper, A., Falcao V., Gibbons J. The Active Badge Location System: ACM Transactions on Information Systems. 1992 10(1). p91–102.

15. Harter A., Hopper A. A New Location Technique for the Active Office. IEEE Personal Communications; 1997 4(5). p43–47.

16. Nissanka BP., Chakraborty A., Balakrishnan A. The Cricket Location-Support System: Proceedings of ACM International Conference on Mobile Computing and Networking (MoBiCOM), 6 – 11 August, 2000, Boston MA USA. 2000. p32-43.

17. Bahl P., Padmanabhan V. RADAR: An In-Building RF-based User Location and Tracking System: Proceedings of IEEE The Conference on Computer Communications (INFOCOM), 26 - 30 March, 2000, Tel-Aviv, Israel. 2000. p775-784.

18. Konrad L., Welsh M. MoteTrack: A Robust, Decentralized Approach to RF-Based Location Tracking: Proceedings of the International Workshop on Location and Context-Awareness at Pervasive, 12 – 13 May 2005, Oberpfaffenhofen near Munich, Germany. 2005. p489-503.

19. Hightower J., Borriello G., Want R. SpotON: An Indoor 3-D Location Sensing Technology Based on RF Signal Strength. University of Washington, Seattle, Tech. Rep. 2000-02-02, February 2000.

20. Castro P., Chiu P., Rremenek T., Muntz RR. A Probabilistic Room Location Service for Wireless Networked Environments: Proceedings of ACM International Conference on Ubiquitous Computing (UBICOMP), 30 September – 2 October 2, 2001, Atlanta, Georgia USA; 2001. p18-34.

9

SEAMLESS HEALTHCARE SYSTEM FOR THE ELDERLY BASED ON RFID TECHNOLOGY

Lan Mu[1], Xiaodan Han[1], Yang Chen[1], Xinzheng Jin[2*]

[1]Huazhong University of Science and Technology, Wuhan, China

[2]Huazhong University of Science and Technology, Center for Health Information Resource, Wuhan, China

ABSTRACT

Purpose: To design a seamless healthcare system to provide the elderly in the community of Wuhan city with active, real-time and seamless health services. Method: All of the health services for the elderly in the community of Wuhan city, such as chronic disease management, first aid, and diagnosis have been systematically integrated based on the Radio Frequency Identification (RFID) technology combined with the existing high-tech, such as nano-sensor, network, information system technology, wireless technology, Bluetooth and Area Position System (APS) positioning. Results: The system has been constructed and put into use, initially achieving seamless health services for the elderly. The majority of both the elderly and healthcare workers are satisfied with the system. Conclusion: The study initially proposes the use of RFID technology to plan and design the elderly healthcare system, and has successfully carried the system out and got positive outcomes. It provides a new direction in the relevant research fields.

KEYWORDS

RFID, The Elderly, Health Services, Seamless Healthcare System

1. INTRODUCTION

The world is faced with the problem of aging. It is particularly notable in China. The Population Division of the Department of Economic and Social Affairs in the United Nations published the "World Population Prospects: The 2010 Revision" [1] , which pointed out that the problem of aging will be even more serious while average life expectancy increasing and fertility rate declining, especially in low-fertility countries. By 2100, aging will continue, the percent of population over 65 years of age will be more than 28%, which means, one of every four people is over 65 years old. According to the "Statistical Communiqué of the People's Republic of China on the 2012 National Economic and Social Development" [2] published by National Bureau of Statistics of China in February 2013, the population over 60 years old will be up to 194 million at the end of 2012, it is 14.3% of the total population.

In china, patients often have to queue for long time and beat up and down among different departments in hospitals. What's more, the cost is much higher in hospitals than in community healthcare centers. Therefore, due to both economic and psychological reasons, many old people who suffer from chronic diseases prefer to receiving home rehabilitation, regularly checking in community healthcare centers to specialized treatment and care in hospitals, which contributes to the use of recourse from the respective of medical resource conservation. But on the other hand, it increases the risk for the elderly.

For the elderly who get rehabilitation treat at home, there is no real-time observation from professional medical staff. It is also very difficult for their family members to take care of them at anytime and anyplace. So if some unexpected accident occurs, such as a sudden illness repeated, stroke, and syncope and so on, and the elderly at home alone is not able to call for help in time, dire consequences, and even loss of life would be followed. The life of the elderly in China and the whole world is being threatened by these emergency situations, and the current clinical technology could not prevent these unexpected situations in advance. So, to construct a system to monitor the high-risk factors in real-time, to provide self-help settings, and to locate the elderly will be a very effective method to solve these problem.

2. THE MEANING OF CONSTRUCTING THE SEAMLESS ELDERLY HEALTHCARE SYSTEM

When a patient is transported from a community healthcare center to a hospital, the health information can hardly be shared or exchanged completely and immediately. The hospital or emergency center could not quickly learn such basic information as name, blood type, allergy, medical history and so on. If the patient is unconscious, having amnesia or Alzheimer, they could not immediately notify the patient's families. These problems will influence the conduct of effective emergency work. All these issues lead to an enormous waste of rescue time and affect the efficiency of treatment, and seriously threaten the life and health of the elderly. They are also currently a huge obstacle to improving the elderly healthcare system. Therefore, the study takes Radio Frequency Identification (RFID) technology as a base combined with network technology, information system technology, nano-sensor, APS and Bluetooth to construct a seamless elderly healthcare system, which cover family, community healthcare centers, emergency centers and hospitals, to ensure that the health status of the elderly can be real-time monitored, managed, and that they are able to receive timely medical assistance. Referrals can be accessible among medical institutions.

3. RFID

3.1. The RFID

RFID is used to automatically identify and monitor objects. RFID system generally consists of electronic tag, reader and antenna. Tag stores data in agreed format. It can be implanted under the skin and can also be made into the shape easy to carry, such as card, watch and lanyard. Reader can be either mobile or fixed, send the monitoring data from the tags to data terminal or data processing system through a wireless or wired network or Bluetooth. Antenna is used for transmitting radio frequency signals between tag and reader.

The basic principle of how the RFID technology work is making use of the transmission characteristics between radio frequency signal and spatial coupling or radar reflector to achieve data's transfer to between objects those can be identified. Depending on the different way of power supply, electronic tags can be classified into two categories. They are active tag and passive tag. Active tag carries with internal battery so it can provide energy itself and automatically sends radio frequency signals to the reader. On the other hand, the passive tag cannot send signals to reader firstly. It needs to receive radio frequency signals emitted by the reader and then sends out the information stored in the chip by the energy obtained from induced current.

3.2. The Use of RFID in Medical Field

Compared with the IC card or barcode technology, RFID technology has more advantages in the health system: 1) Readers can automatically identify the tags and conduct read and write operation without contact of tags; 2) Readers could identify the tags remotely and it is also be able to automatically identify tag moving at high speed; 3) Readers can simultaneously read and write multiple RFID tags within range and there is no interfere with each other; 4) The storage capacity of the tag can reach 128 k or more, which makes it capable of storing more information; 5) Tag can work in extremely harsh environments. It has great resistance to high temperature, tempering, acid, base, etc. [3] ; 6) The intrinsic electromagnetic radiation will do no harm to people.

4. THE CONSTRUCTION OF THE SEAMLESS ELDERLY HEALTHCARE SYSTEM

4.1. RFID Smart Wristband

The basis of this system is to provide each old people with chronic disease or with high risk factors with RFID smart wristband. The features of the smart wristband are as the following:

1) Identification

Identification is the most basic function of the RFID smart wristband. Electronic tags store the basic information of the elderly, such as name, age, identity card number, home address, contact information, etc. Readers can automatically identify the information stored in the tags.

2) Storing electronic health records

Electronic tags store the important personal electronic health record information, especially the patient's health history, drug allergies, medication, necessary medical information, etc.

3) Seal-time monitoring of the physiological state of the elderly

Tags in the RFID smart wristband monitor the main vital signs of the elderly and the status of the main parts of the body by nano-sensor. The readers will regularly read and write real-time monitoring information. Once detect anomalies or some index exceeds a preset threshold range, the smart wristband will automatically send warning beep and will send the alarm information to the monitoring terminal.

4) Real-time location

The reach range of RFID readers need to cover all corners of the communities or hospitals. RFID readers automatically identify the smart wristbands and the monitoring information in it to monitor terminals to obtain its current location and achieve real-time positioning in small range. Besides, we may also adopt Area Position System (APS) which uses active RFID technology. APS is capable of timely and accurately reflecting the dynamics of the elderly to the monitoring terminals in the community center or hospitals.

5) Quickly seek for help

A self-alarm button is set in the RFID smart wristband. Therefore, once an emergency occurs, the smart wristband will transmit the alarm information to the monitoring terminals or emergencies so that the elderly could quickly seek for help in critical situations.

4.2. The Design of the System

The system can be divided into three levels. They are data collection layer, the data transport layer and data application layer. Data collection layer is mainly composed of an antenna, a reader, a portable RFID read and write device (referred to as PDA). This layer is the basic layer of the entire system and its main function is to identify and read all kinds of data in the electronic tags of the RFID smart wristband worn by the elderly. The data transport layer mainly means that the data are transmitted through the network. It plays as a bridge that links the data collection layer and data application layer. The data application layer means devices which take use of the data in the RFID tags, including monitoring terminals, electronic bulletin boards, PDA handheld by medical personnel, the database server and the server of hospital information system (HIS). Among them, the core database and server of APS real-time location and tracking system are embedded in both community database server and HIS serve. Both monitoring terminals and PDA of medical personnel can be used as the application services terminal of APS to achieve the real-time location and tracking of the elderly in communities. This system takes HIS as the center, spreading all over community medical institutions and the house of community residents. Its structure is shown in Figure 1.

4.3. The Application of the System

The system has been applied in two communities in Wuhan city. Centered on the old people who wear the RFID smart wristbands, the system monitored the physiological status of the elderly when they were at home, in community healthcare centers, hospitals or other emergency centers, monitored and managed the whole process of medical behaviors. The system used RFID technology to achieve mobile management of the old people, successfully integrated the different stages of life management, chronic disease management and emergency diagnosis to achieve seamless connection both in time and space. We can see hardware structure the system in Figure 2.

5. OUTCOMES

After the system being put into use, we mainly have achieved the following goals.

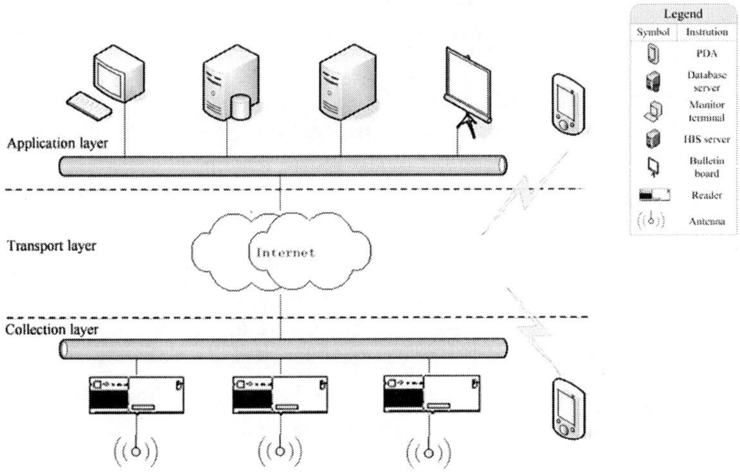

Figure 1. The system structure of seamless elderly healthcare system based on RFID.

5.1. In Family and Community-Life Management

This is mainly for family which the other family members don't live with the old together so the old people live alone. The RFID smart wristbands will monitor the physiological status of the elderly and the health information will be sent to the readers configured for family, and then periodically be sent to the families of the elderly through Short Messaging Service (SMS) or the network to help family members know the health condition of the elderly they care for as soon as possible and whether there is an abnormal condition.

Meanwhile, the readers also regularly send the physiological status information to monitoring terminals in community through a LAN or Bluetooth. So the monitor stuff will know the health status of the elderly living alone in real-time. Once an unfortunate fall or sudden illness cause alarm or the elderly self start the help devices, RFID smart wristbands will quickly send out radio signals and the location of the elderly to alert the readers to convey the information to monitoring terminals and the families immediately.

5.2. Community Healthcare Centers-Chronic Disease Management

Community healthcare centers should actively provide service for the elderly with chronic diseases. Community healthcare centers get the physiological information of the elderly with chronic diseases from the community monitoring database. So they could observe whether the health status of the elderly has been improved or not and provide further reference to medical personnel to make diagnosis and take treatment. In addition, community healthcare centers should regularly provide the elderly with free examination and lectures about chronic diseases management-related knowledge. All of the progression of the disease and the treatment of the elderly must be recorded in RFID smart wristband to form a complete and continuous electronic health records.

5.3. Hospitals or Emergency Centers-Emergency Diagnosis and Treatment Management

Elderly patients community healthcare centers could not treat can be referred to a higher level hospital and in case of an emergency situation can be transferred to the emergency center. The PDA configured in ambulance can read basic information, family contact information, health history information and current physiological status values of the elderly from the RFID smart wristband, which can help the ambulance crew take more accurate and effective emergency diagnosis and treatment and contact the families of the elderly quickly.

Besides, readers should be installed all over the hospitals or emergency centers so that the monitoring terminals could identify and locate the elderly wearing a smart wristband in real-time.

The location health status information can be published in the electronic bulletin board to make sure that the doctor and families of the elderly know his location and physical condition at any time, reducing the families' fear and preventing lost, falls and other accidents from happening. The PDA of the staff of hospitals or emergency centers can help them identify the patient accurately and further prevent and reduce medical errors caused by patients' identification errors. There are so much information such as allergies history, health history, and test results, etc. that can be read from the RFID smart wristbands worn by the elderly. This can help save costs and avoid duplication test and even avoid the possibility that the elderly may deliver wrong message to the doctor due to memory loss or other reasons.

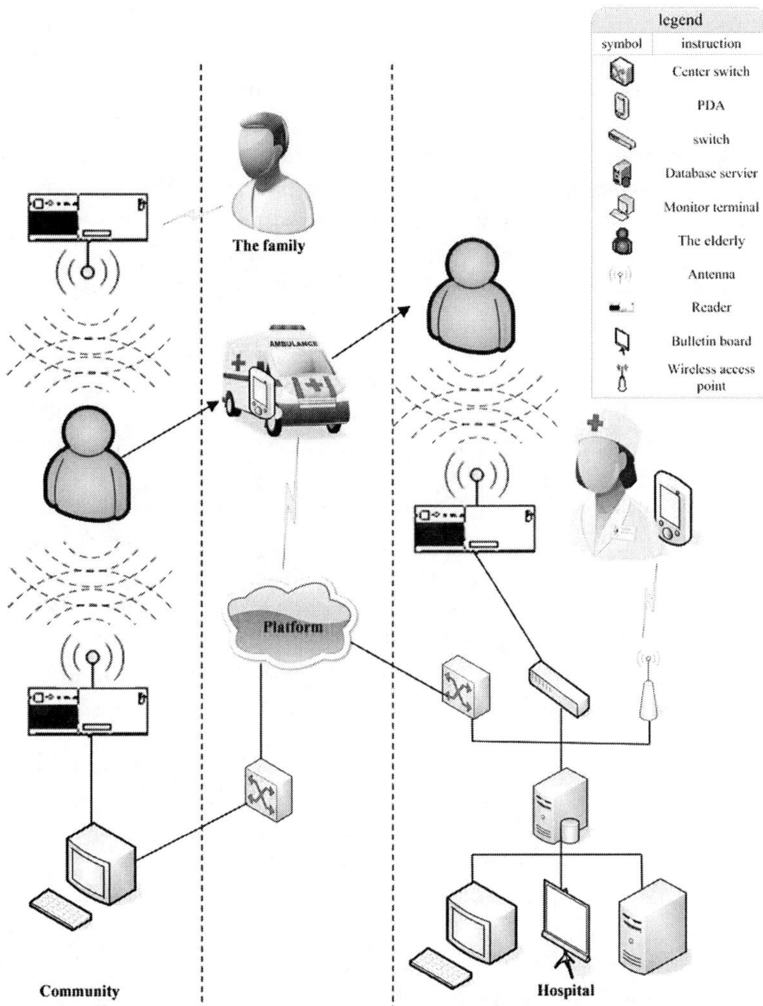

Figure 2. The hardware structure of seamless elderly healthcare system based on RFID.

6. DISCUSSION

6.1. Comparison with the Traditional Healthcare System for the Elderly

Traditional healthcare system also includes home, primary healthcare agency, hospital and emergency center. But the main way how information is transmitted is quite different from this new system. In the traditional way, the delivery of information mainly rely on oral or notes. In this way, the transmission of information is not only of low efficiency, but also is the cause of many other problems. It may delay the treatment. Error in expression, misunderstanding, a slip of the tongue or a slip of the pen and so on, which does most likely happen in oral or notes. All of these problems are likely to lead to improper treatment or medical errors.

In addition, there is no sharing and common platform among medical institutions in the traditional system. The information of various medical institutions is relative independent so that medical institutions cannot share or exchange information or can just share a little. Bad information sharing has seriously affected the effective use of information resources. Moreover, the traditional healthcare system is lack of spiritual of initiative to providing healthcare. It just responds to seek for medical services proposed by the elderly. Its branches agencies usually scatter at various points. So the elderly have to walk around to get help, which adds to the danger while the elderly may fall ill on their way.

The seamless elderly healthcare system is based on RFID technology. The RFID smart wristband worn by the elderly plays a core role in the process of health management. It is throughout the daily lives and medical behaviors of the elderly. It has excellent memory and reacts at a very high speed. It seems like that the elderly have a 24 hours 7 days uninterrupted personal nanny. This system has brought great benefits to both the patients and medical institutions: 1) The system makes both the elderly people living alone and their family members get a stronger sense of security; 2) It eases the complexity of the elderly going for medical services; 3) It reduces the incidence of medical accidents or errors; 4) More importantly, it has improved efficiency of emergency treatment for the elderly. More accurate and adequate information save more time for rescue; 5) The system makes it possible for multi medical institutions to share and exchange information through the common platform, so that information sharing and exchange between medical institutions could be enhanced; 6) All data in this system are stored in the database server in the monitoring terminals. Any communities, hospitals or emergency centers can use the information as needed. The RFID common platform reduces multi-point information storage and information store disorder, avoiding obstacle of information use and the waste of information resources.

6.2. The Evaluation of the System

6.2.1. The Design of the Survey

We interviewed the elderly and their family members to survey the changes in their satisfaction and medical expenses before and after using the Seamless Healthcare System. Simultaneously, we also visited healthcare workers from community health centers and hospitals to survey their satisfaction to the system. From Table 1, this survey included totally 948 respondents, including 528 male and 420 female, who are all from two communities of Wuhan city. 495 respondents are the elderly and 397 are their family members. The other 56 respondents are healthcare workers, 28 from community health centers and 28 from hospitals and emergency centers.

6.2.2. Outcomes

Totally, as Table 2 shows, 86.0 percent of the respondents were satisfied with the system and agreed with that the system had improved their quality of life after using the system, while 12.8 percent thought that there was no difference before and after using the system. In addition, a minority (1.2%) were dissatisfied with the system, for their cost in network fee had increased. The respondents' satisfaction degree increased significantly after the implementation of the system ($p < 0.01$). The average medical costs of the elderly didn't change significantly within the first year, but decreased on average in two or more years (see Figure 3). The main reason is that the equipment (such as the wristband) occupied a certain proportion of medical costs at the first year. From Figure 4, we could see that the majority of healthcare workers agreed with that the system had helped them improve the efficiency of their work, while staffs from community health centers had higher satisfaction.

Table 1. The testing example.

The composition of respondent					
Residents		Healthcare workers		Total	
The elderly	Family members	Health centers	Hospitals and emergency centers	Male	Female
495	397	28	28	528	420

Table 2. Respondents' satisfaction to the system.

			Satisfaction classification			Statistics	Asymp. Sig. (2 tailed)
			Dissatisfied	No difference	Satisfied		
Group	Before intervention	Count	53	406	489	Z = −16.072	p = 0.000
		% within Group	5.6%	42.8%	51.6%		
	After intervention	Count	12	121	815		
		% within Group	1.2%	12.8%	86.0%		

Note: Significant at $p < 0.01$.

In general, after the implementation of the seamless healthcare system, the elderly and their families' satisfaction degree has improved. Although medical costs can not reflect its advantage immediately, it can reduce medical costs in the long term, which has economic benefits.

6.3. Shortcomings

We find some problems that need to be improved in the application process of this system.

6.3.1. Security

The biggest challenge faced by the seamless elderly healthcare system is the issue of privacy and further security issues arising from privacy. In the RFID system, the radio waves can pass through the building and metal, and are non-contact communication between the reader and the tag, so the channel eavesdropping is possible and can even be very serious.

The limited storage capacity of the electronic tag limits the complexity of both the encryption algorithm and access process. These factors are all vulnerabilities which may be taken advantage by information thefts who can steal a large amount of private health information from this system. The privacy of those who use the electronic tag is hard to be ensured and the security and privacy of users are under threat.

6.3.2. Privacy

In the seamless elderly healthcare system, the information in the RFID smart wristbands is related to many individuals and institutions. It is the base of most of the medical behaviors. If we simply encrypt the data in the electronic tag, it is most likely to affect the transmission of the data. For instance, emergency personnel may not obtain the basic information and medical information of elderly for they don't know the secret key of the encryption algorithm. The emergency treatment of the elderly may be possibly delayed.

Currently, there are many existed solutions to eliminate conflicts between the application of RFID technology and privacy protection, such as active signal shielding, active signal interference, encryption technology, Hash switch, Kill commands, physical unclonable function (PUF) and linear feedback shift register (LFSR) protocol, etc. However, all of the application of these technologies in the RFID system has some flaws and cannot fully adapt to the complexity of the present system. Therefore, to solve the privacy problem the seamless elderly healthcare system faced, we need to continue to study the issue in depth to make sure that the system can be implemented smoothly.

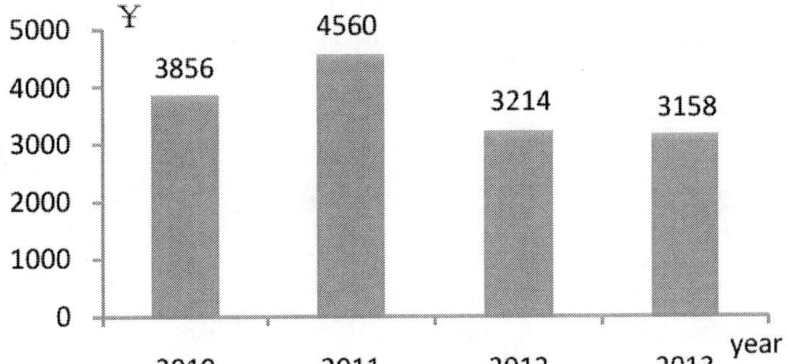

Figure 3. The average medical costs of the elderly.

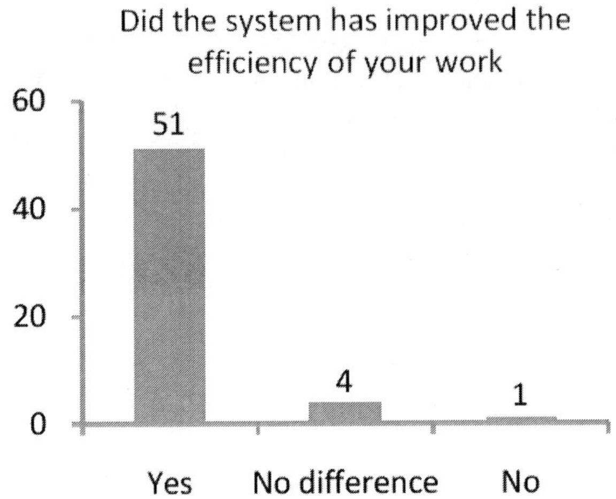

Figure 4. Healthcare workers' attitude to the system.

7. CONCLUSIONS

This study is the first practice of integration of the application of RFID technology and the elderly healthcare system. The seamless elderly healthcare system has been applied in two communities in Wuhan city, China. It has successfully improved the quality of medical the elderly in the community have got and saved the cost of medical care. This expands the new research directions for research in related fields.

However, because there are many differences between the communities, further study and discussion is needed for seamless elderly healthcare system based on RFID technology. In addition, RFID security and privacy issues also need to be further improved.

REFERENCES

1. National Bureau of Statistics of China (2013) World Population Prospects: The 2010 Revision. United Nations Department of Economic and Social Affairs, New York.

2. National Bureau of Statistics of China (2012) Statistical Communiqué of the People's Republic of China on the 2012 National Economic and Social Development. National Bureau of Statistics of China, Beijing.http://www.stats.gov.cn/english/StatisticalComm-uniqu201302/t20130222_61455.html

3. Zhang, Z.B. (2007) The Application of Radio Frequency Identification (RFID) in Healthcare Industry. Medical Equipment, 1, 3-5.

10

A COMPUTATIONAL ARCHITECTURE BASED ON RFID SENSORS FOR TRACEABILITY IN SMART CITIES

Higinio Mora-Mora *, Virgilio Gilart-Iglesias, David Gil and Alejandro Sirvent-Llamas

Specialized Processors Architecture Laboratory, Department of Computer Science Technology and Computation, University of Alicante, 03690 Alicante, Spain

ABSTRACT

Information Technology and Communications (ICT) is presented as the main element in order to achieve more efficient and sustainable city resource management, while making sure that the needs of the citizens to improve their quality of life are satisfied. A key element will be the creation of new systems that allow the acquisition of context information, automatically and transparently, in order to provide it to decision support systems. In this paper, we present a novel distributed system for obtaining, representing and providing the flow and movement of people in densely populated geographical areas. In order to accomplish these tasks, we propose the design of a smart sensor network based on RFID communication technologies, reliability patterns and integration techniques. Contrary to other proposals, this system represents a comprehensive solution that permits the acquisition of user information in a transparent and reliable way in a non-controlled and heterogeneous environment. This knowledge will be useful in moving towards the design of smart cities in which decision support on transport strategies, business evaluation or initiatives in the tourism sector will be supported by real relevant information. As a final result, a case study will be presented which will allow the validation of the proposal.

KEYWORDS

RFID smart sensor network; cyber physical systems; communication technology; computational architectures; smart city; track and trace system

1. INTRODUCTION

Technological progress in recent years has made it possible to extend the use of Information Technology and Communications (ICT) to new applications with the objective of improving citizens' quality of life. This idea has been of increasing significance in the political agenda as well as in the public services area. In the cases where there is no planning of development consistent with the efficient and sustainable management of urban resources this can start to cause real issues.

It is recognized that to either obtain a common definition of Smart City (SC) or to define global trends for SC is complicated. There are several reasons that contribute to this. On one hand, cities are complex systems typified by massive numbers of interconnected citizens, a lot of businesses, various modes of transport, communication networks, including services as well as utilities. On the other hand, this leads to a big diversity of technological, social, economic and organisational problems which expose the environmental sustainability of cities. Besides, there is no total consensus about the significance of SC but there is wide agreement concerning the fact that SCs are characterised by the use of ICTs. The goal is, in general, to make better use of resources in cities [1]. There are numerous examples which we could cite of cities that are supplementarily prepared with ICT systems in order to meet this objective in the fields of energy, telecommunications, ecology or transport [2,3,4,5]. Improvements in the management efficiency of these elements are directly perceived by citizens as quality of life improvements.

These reasons have led us to work on the citizen track and trace problem in urban environments. Progress on this issue may provide relevant information for decision-making in many of the above problems related to the management of a smart city, such as in the planning of transport infrastructure (roads, subway, bus lines) or modification of existing ones, in the construction of new public buildings (schools, hospitals, museums, sports areas), in designing tourist routes through the city, *etc.* Other uses of the obtained information can contemplate economic objectives such as a commercial rating of avenues and streets or the design of tourist offerings in the city.

Moreover, building increased intelligence into things around us and connecting them to the network to transfer data without requiring human interaction is a current trend of technological development called the Internet of Things (IoT) that allows us to meet and interact better with the surrounding environment [6,7]. IoT technologies such as sensing, computing and mobile communication could be employed to collect data on the functioning of the city. In this respect, how the data of interest is collected is the major challenge in most cases. Once obtained, another important issue is how to connect sensors to the web [8] to enable provision of cloud-based management and services. In this field, the development of distributed sensor infrastructure is one of the prominent research areas related to IoT [9,10].

In the work described in this paper, a method and a computer architecture to determine the movements of people in the city is proposed. The research done takes into account the current communication technologies, the ease of implementation and deployment and key aspects about user privacy. The main scientific contribution of this research is the integration of several new technologies and solutions of IoT to build a comprehensive system to track and trace citizens in city working environments. As a result, smart sensing technologies with processing and communication capabilities are proposed to produce cloud services for citizen traceability.

The remainder of the paper is organized as follows: Section 2 covers problem analysis, related work and the research methodology used; Section 3 presents the methods to know citizens' movements with its dominant distributed components and decisions taken; Section 4 discusses aspects of communication and structuring of location information and describes the centralized architecture; Section 5 describes the main characteristics of the services provided for using the generated information; next, Section 6 describes the implementation based on a case study in a university environment to check the correct performance; finally Section 7 concludes with a discussion of critical outstanding issues and future research directions in this area.

2. POPULATION TRACK AND TRACE ISSUES

2.1. Problem Definition

Track and trace is a term frequently used by delivery and logistics companies [11,12,13,14,15]. In the context of items and products, track and trace implies a process of determining their traceability understood as the current and past locations (or even more information) of a single item. In this work, the main idea is to use this general approach in a totally different context regarding people in order to discover and determine citizen movement flows. According to this definition, the general problem addressed in this paper focuses on achieving structured information about the movement habits of citizens in the geographical area of a city.

The main challenge to be resolved is the way in which the population can be traced preserving their anonymity, without necessarily requiring their collaboration, and in non-controlled and heterogeneous environments. In addition, the processing and centralization of information obtained must be considered to be useful for decision-making, and it should be aware that the number of citizens can be very large, even considering a small urban geographical area of an average city.

2.2. Literature Survey

The citizen track and trace issue is not new, as evidenced by the amount of research on this topic. Indeed, there is some concern about maintaining user privacy due to the proliferation of systems and high integration rate of mobile devices connected in the hands of citizens [16,17]. However, the benefit of knowing citizen's traceability information for the design of cities has spurred studies and works which try to determine it while preserving the privacy of citizens.

Transport planning is an example of a task where an analysis of human and traffic flows can be very useful. This issue is essential since the populations of cities are increasing dramatically. Decisions regarding this have been studied for many years. As a result, there are numerous proposals for successful decision-making processes in transport planning [18,19,20,21]. Such decisions involve many actors such as members of the city council, representatives of the transport companies and even the citizens. Usually, council members along with the company representatives come together to form a regional consortium. Therefore, this consortium is the scenario where the decisions regarding public transportation are made. Setting the routes of the various services such as buses or subways is one the main decisions they have to perform. Due to this fact, they must analyse data about the city and its citizens in order to extract the knowledge about their habits and transportation necessities, *i.e.*, the analysis of human movement trends in urban areas. However, although these guidelines provide a well-studied framework for decision-making on transport, there are some critical shortcomings that have been detected in some of the current literature [22,23].

First of all, very often there is an issue of the obsolescence of design plans. Traditional planning processes can become obsolete when cities grow significantly. If too much time passes between when a project is planned and it is executed, by then it will be outdated and certainly will have lost some of its usefulness. To address this problem, there are works proposing a system based on the use of SC tools that allow projects to evolve dynamically with the possible changes in the scenarios over time [24].

Secondly, there is another important issue concerning the procedures used to get the data and especially how can they be used for providing decision support. Traditionally, the information used in these processes is obtained through field studies (e.g., manual recounts of people and vehicles, census, *etc.*) along with information obtained from interviews with the different actors involved, namely, citizens, transportation companies, and event managers, among others [22]. However, these massive surveys and interviews are very expensive, thus they cannot be performed frequently. Due to this fact, they rapidly become obsolete and if we trust such outdated information we will not take into account the real situation. Furthermore, these data sources lack information about individual trajectories and consequently we cannot compute the real individual movements of people [22,23], but rather we are able to obtain a global and a simple view of citizens' movements.

It is highly desirable, therefore, to use technology to effectively determine the human traffic flows. The main technology created for this purpose is the Global Positioning System (GPS) [25]. With GPS, we can precisely and easily establish our position on the world map, so it is used for orientation purposes by drivers and pedestrians since it has become quite affordable. In addition, it is now integrated into many devices such as mobile phones or other ubiquitous devices [26].

An example of an application for this problem can be found in a study conducted in the cities of Milan (Italy) [27] and Beijing (China) [28]. In the case of the Italian city, the data come from GPS devices installed in a rental car fleet. In the Chinese city, the GPS devices were installed in a taxi fleet. These papers present some analytical methods as well as data mining tools in order to provide discovery knowledge processes. They rely on a central node responsible for processing all the information acquired from the GPS devices.

Based on the information obtained about the trajectories, there are several studies which focus on data analysis and manipulation to extract information of interest. In this regard, query languages and tools for trajectory analysis oriented to query trajectories computed from GPS devices have been established [29]. These researches develop a clustering technique that allows extracting traffic patterns over indicated periods of time. The results of these analyses allow one to trace road communication routes much faster than existing patterns and the results obtained by tools like Google or Bing maps.

However, the use of GPS raises some issues to consider. One of the most important of them is the transparency of use: access to the positioning data of users requires that they allow it, either by configuring or installing an application designed to carry out such data collection on their mobile devices. This configuration involves extra effort that not everyone may be willing or able to undertake. Moreover, if it is activated, the user will be aware that is transmitting data at all times. Therefore, it becomes clear that it is not a transparent mechanism for the user. In addition, there is an added problem: the preservation of privacy [30]. By knowing the exact location of people in real-time and the trajectories they have followed over time, it is possible to study their movement habits and even predict their location at a certain time of day. In the wrong hands, such information may compromise not only the privacy of citizens, but also their safety. For instance, it would be possible to know with precision at what a time a citizen is not at home in order to commit a theft. Finally, besides the two mentioned problems, an additional problem is the power consumption since devices require their own power source to operate in addition to the cost of the devices themselves. Despite being integrated into many medium and high range mobile phones, they are still relatively expensive if purchased separately.

The use of the installed wireless infrastructure, has also been considered for this application [31]. This includes mobile phone networks and wireless local (Wi-Fi) and wide (WiMax) area networks. These networks were not designed for this purpose but to rather access the Internet and communications services offered by telcos. However, from the moment in which each base station is aware of the presence and identity of each connected device, it is possible to offer tracking operations while the user is moving from one station to another [32,33]. Some research also includes filtering and interpolation operations to predict where the user is located [34,35,36]. However, due to their scope and current coverage, they cannot provide accurate situation information about where the user has passed. Moreover, placing a greater number of stations for this purpose could significantly increase the infrastructure cost.

Mobile telecommunications technologies (GSM, 3G, LTE, *etc.*), can also be used to perform large-scale tracking and tracing of users, however, they are not considered in this work due to the lack of necessary precision and the need of having a mobile terminal. Although their degree of penetration in the population is very high, their use for this purpose involves other drawbacks about cost and user privacy. Other alternatives such as CCTV, are discarded due to the technical difficulty of developing automatic methods for tracking individual users in uncontrolled contexts and the fact it can cause a stronger invasion of privacy due to the feeling of being "watched" by surveillance cameras.

An alternative to previous techniques for sensorization of citizen movements is to use Radio Frequency Identification (RFID) communication technology [37,38,39]. This technology has also been widely used to obtain location information of users and objects. Table 1 summarizes the main recent works on this issue as a representative sample. All the proposals analyzed work in controlled environments, that is, the infrastructure deployment has been designed previously and provided by the system managers. When the users access a system are given a tag to carry around the working environment.

Many studies have been conducted for track and trace indoors and outdoors. For instance, outdoors, the City of London introduced a system for data collection in its transportation system including buses, subways and trains [40]. The idea was to track users of the transport network through payment and access cards equipped with this communication technology. With the information gathered it was determined the level of use, number of travelers and public transport habits in the city and to know the stations, breakpoints, origin and destination of network travel streams. For indoors, systems to know citizen movement flows have also been implemented [41], for instance, visits per museum [42]. With the results, visit patterns of the different exhibits and areas were obtained, participants with atypical behaviours were identified and even relationships between visitors groups were established.

RFID technology overcomes some of the disadvantages described for GPS, as it does not require too much cooperation from the user and reduces energy costs. However, its use for tracking does not provide exact locations but, like other wireless networks, the temporal location marks are limited by the locations of the signal reader antennas and the analysis of the Received Signal Strength Indicator (RSSI) of the readings [56]. Thereby the resolution of the trajectories followed by the users will be lower than in the case of the GPS. The outdoor scope of this technology is also not comparable to GPS coverage. There is a shorter range variation named Near-Field Communication (NFC) [59] that requires much less power to operate. Moreover, passive RFID/NFC technologies may be present in a multitude of objects and devices that people wear [38]. In many cases, they are integrated into standard cards made of plastic with identification and access functions [60], credit or debit cards [61,62,63] or even paper tags on clothing [64,65,66,67], however the functions for which these elements were designed were not tracking people in itself, but rather other applications such as localized identification, payment processes or object tracking in stores.

After the review carried out, it is observed that the activities of localization and tracking can provide valuable information to improve infrastructure development and user satisfaction for a variety of applications. From all the reviewed literature on track and trace systems, the most cited solutions show significant drawbacks. The Global Positioning System is the most suitable and it provides better results due to its precision and accuracy, however, this implies the necessary participation of citizens for track and trace activities. The citizen must have a device with computing and communication capability that enables it to provide the captured information, usually a mobile phone on which an application has been previously installed and configured. These requirements consume a considerable amount of energy that may affect the autonomy and the availability of the device. In addition, these solutions involve knowledge of personal data of the citizen like the terminal number. These features imply that the system is restricted to a limited set of the citizens, so the sample would be not representative to provide good service for managing the resources of all citizens. Elderly citizens who are not familiar with ICT, citizens who do not have access to such technology or citizens unwilling to provide personal data, among others, would not be represented in the datasets. Moreover, these proposals are only valid for open city environments, which would imply the inability to track citizens in indoor environments where GPS technology is not adequate. Nevertheless, these spaces also require efficient management of their resources.

To take advantage of the extensive set of connected mobile devices as well as the installed infrastructure, the use of wireless networks is another option in which proposals have already been made. However, the results it gets are not very precise and it also produces privacy intrusions as the identification of the user is known.

Another alternative that can be provided is the use of RFID/NFC, a technology increasingly found embedded in devices and other portable fixtures accepted by the public. For its functioning, not only energy consumption, but also expensive equipment, are not required. In this case the main obstacle is that the technology is underdeveloped and in general, antenna networks prepared for this purpose are not available.

Most related works that use RFID technologies are focused on tracking systems and only few proposals have contemplated both the track and trace problems. Most of them apply their solutions in closed and limited environments, where the characteristics of reliability, service quality and scalability are not necessary. However, the development and proliferation of this RFID technology among citizens during the last few years, opens up new possibilities for its use in track and/or trace tasks based on user-centric schemes. The characteristics of this technology may allow mixed deployment both indoors and outdoors. In the analysed work, the use of RFID technology is limited to reading the homogeneous tags provided by the system owner and sending information to a central system, without prior processing, storage or filtering, as these tasks are delegated to a central node, which can generate a bottleneck.

It is noteworthy that none of the literature proposals focus on the provision of the information collected to third parties, in a dynamic, personalized and open way. For this purpose, the development of cloud-based architectures may represent a good choice for the design of flexible management architectures and service providers. This aspect could enhance the development of decision support systems for the efficient management of resources and oriented to meet the needs of citizens. The scope of the limitations and advantages mentioned will be discussed in the following sections to lead to the design of the proposed architecture according to the objectives.

Table 1. RFID-based track and trace works.

Work	Main Application	Working Environment		Tag Deployment	Reader Deployment
		Indoors	Outdoors		
RFID enabled supply chains [11–14]	Traceability services	X	X	Passive tags embedded into objects	Fixed or mobile readers. Along supply chain
RFID delivery system [15]	Intelligent Transport Systems		X	Passive tags embedded into objects	Fixed or mobile readers. Along delivery system.
RTSV [43]	Tracking System for Vehicles		X	Passive tags installed on cars	Distributed throughout the city.
The London Oyster Card Data [40]	Public transport planning	X	X	Passive tags Inside transport cards	Installed on entries and exits of transport system.
iWalker [44]	Assistance services of location and obstacle detection	X		Placed anywhere in the environment	Embedded into walkers
Tracking science museum [42]	People Traceability	X		Passive tags on nameplates carried by users	Distributed throughout the floor
SIP-RLTS [45]	Location Tracking System	X		Passive tags carried by users (patients)	Readers carried by workers (medical)
LANDMARC [41,46]	Indoor location sensing	X		Active tags on grid array deployment	Distributed throughout the environment
Blind User [47]	Location and Proximity Sensing	X	X	Passive tags; Indoor: Grid array deployment over floor; Outdoor: Along edge of the sidewalk.	Readers carried by users
Pervasive mining [48]	Tracking people in a pervasive mining environment.	X		Passive tags carried by users	Distributed throughout the environment
Privacy-Preserving Solution [49]	Tracking People in Critical Environments	X	X	Passive tags carried by users	Distributed throughout the environment
Social interaction [50]	Person tracking	X		Passive tags carried by users	Distributed throughout the environment
Cameras and RFID [51,52]	Tracking and identification people	X		Passive tags carried by users	Distributed throughout the environment
RFID Inside [53]	Tracking and identification people	X	X	Passive tags inserted in users	Readers carried by workers (medical, security, *etc.*)
Tagging Demented Patients [54]	Tracking and identification people	X		Passive tags carried by users	Readers carried by workers (medical)
WSN and RFID [55]	Person tracking	X		Passive tags installed on objects or people	Distributed throughout the environment
Elderly Living Alone [56]	Person tracking	X		Passive tags carried by users	Distributed throughout the environment
Peer-to-Peer Networks [57]	Location Tracking System	X		Active tags carried by users	Distributed throughout the environment
REACT [58]	Children location		X	Passive tags carried by children	Distributed throughout the environment

3. RESEARCH METHODOLOGY

The working hypothesis of this research is that it is possible to know and process citizens' movement flows across the city by using new IoT technologies.

Once the problem has been described, the hypothesis established and the related work analysed, it is necessary that the comprehensive track and trace system becomes clearer. This system should allow the acquisition of locations of citizens in heterogeneous and uncontrolled environments, indoors and outdoors, to facilitate the transparent incorporation of citizens into the system. Furthermore, the information acquired should enable providing information about citizen movement flows to someone else for efficient management of the resources of cities and to be able to provide better services to these citizens.

In order to develop our proposal, a research methodology based on business process management [68,69] using the Eriksson-Penker formal notation [70] it has been followed. Process management is a strategy for structuring a complex process into a sequence of tasks that can be shown as actions that transform inputs into some other output elements. This transformation must be aligned with previously defined goals, which are considered as strategic, to meet some needs or gaps identified in the problem context. In this way, the defined process will achieve the object system of this work in a systematic way, selecting the most appropriate techniques and tools to meet the objectives in order to solve the identified problems. In the proposed methodology, each of the tasks identified represent a stage of the research, and therefore these will have associated one or more scientific methods as described below.

Specifically, from the starting hypothesis, Figure 1 shows the main process carried out in the research, named *Design of citizen track and trace system*, where the input element (*«input»*) represents current track and trace systems. This *«input»* must be transformed through this process into a novel citizen traceability service (*«output»*), addressing the needs and gaps identified previously, which now are represented by the strategic objectives to be able to guide the research process (*«goals»*). To perform the transformation to achieve the objectives established, the controllers (*«Control»*) and facilitators (*«supplies»*) which are required to guide this transformation must be identified. Particularly, these elements represent strategies, paradigms, techniques and technologies to be integrated into our proposal. In order to make easier its identification and, therefore, make clearer the problem, the main process has been divided into three sub processes or tasks.

The resulting sub-processes are the following (Figure 1): first of all, the sub-process *Design of the method for citizens' location acquisition*, focused on establishing necessary technology infrastructures to get citizen data. Secondly, the sub-process *Design of the method of communication and structuring of citizens' location*, involved in ensuring the correct reception of the citizen location data. Finally, the sub-process *Design of the method for citizens' flow provisioning* focused on provisioning standard citizen flows to third parties.

The result of this process will be a computational architecture in order to read and process citizens' movement following the overall scheme shown in Figure 2. The final design of this architecture will be led by the decisions made in each sub process of Figure 1. The research made, uses the knowledge and experience of the research group in the field of designing specialized computational architectures [71,72,73].

As can be seen, it is composed of a distributed module and a centralized one. The distributed module (A) implements the reading of user location data. The centralized module (B) acquires the data from sensors and provide to the consumers (C) the citizen traceability service to help when making decisions on urban management. The functioning and components of those modules will be detailed progressively in the next sections.

3.1. Method for Citizens' Location Acquisition

This task is focused on the identification of the most suitable location technology and the location infrastructure design as a starting point for addressing the problem to achieve the following goals, as noted in Figure 1: gathering of geographic information (*location*) should be transparent to the citizen, meaning without requiring their cooperation and without their intervention in managing and setting the devices, without affecting its economy and independently of whether the citizens' transits through both indoors and outdoors environments (*transparency*); the designed method will have to maintain the user anonymity and protect personal privacy, avoiding any capture of citizens' personal and confidential information (*anonymity*); the technology used should allow its functioning and quality of service (QoS), regardless of the number of citizens interacting with it and the size of the area that will be monitorized (*scalability*) and the associated infrastructure should provide agile adaptation to any environment (*flexibility*); finally, the deployment of the necessary infrastructure should have a low financial and consumption costs (*costs*). With these design goals in mind the intention is to facilitate the placement of a valid infrastructure and minimize citizens' rejection. Figure 3 shows a complete schematic of this process following the methodology criteria.

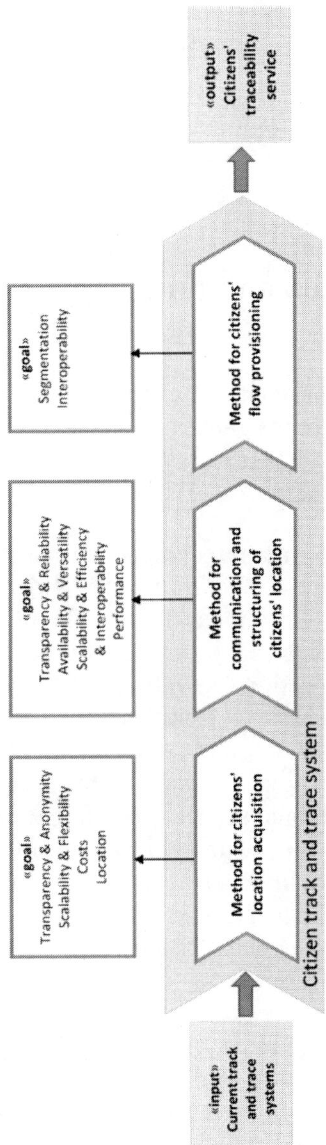

Figure 1. Modelling with Eriksson-Penker notation of the process of obtaining citizen movements flows.

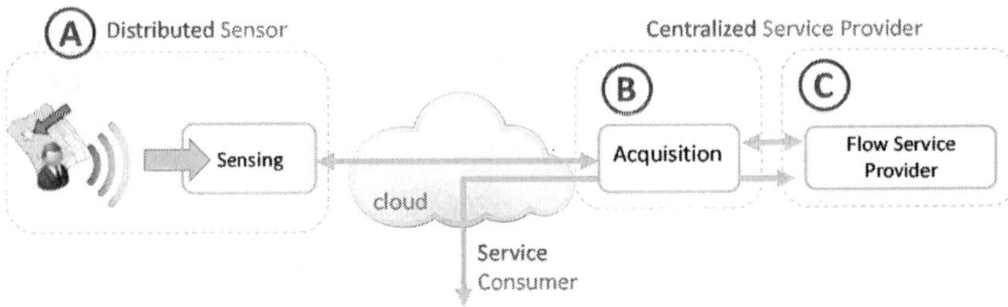

Figure 2. Overall Computational architecture of the citizen track and trace system.

3.1.1. Identification and Analysis of Localization Technologies

For the selection of localization technology, we take into account the analysis and the experience gained in the review of other work and the performance and benefits that are currently available [25,26,30,34,35,36,38,42,59,74,75,76,77].

Table 2 summarizes the main technical characteristics as they relate to the proposed objectives. Regarding the objective of transparency, obtaining positioning with GPS systems or Wi-Fi networks do not offer transparency and the desired anonymity for such applications, since in both cases the collaboration as well as users' permission is required. Specifically, it is necessary that a citizen has a device that supports these technologies, he must install and configure an application designed for this purpose, provide the permission to transmit location information, which may include personal information, and, for all this, he must have a minimum knowledge and technological skill. In addition, the accuracy and coverage of GPS could be a privacy and legality concern, in terms of providing more than the location information necessary for the purpose of the proposal, which is the efficient management of city resources. Besides, although other wireless technologies (Wi-Fi, GSM, *etc.*) could determine the location, the accuracy to address the problem described would not be valid in all cases. All this could hinder the acceptance of the proposal by the general population and therefore it would not be possible to obtain representative samples for the established purpose.

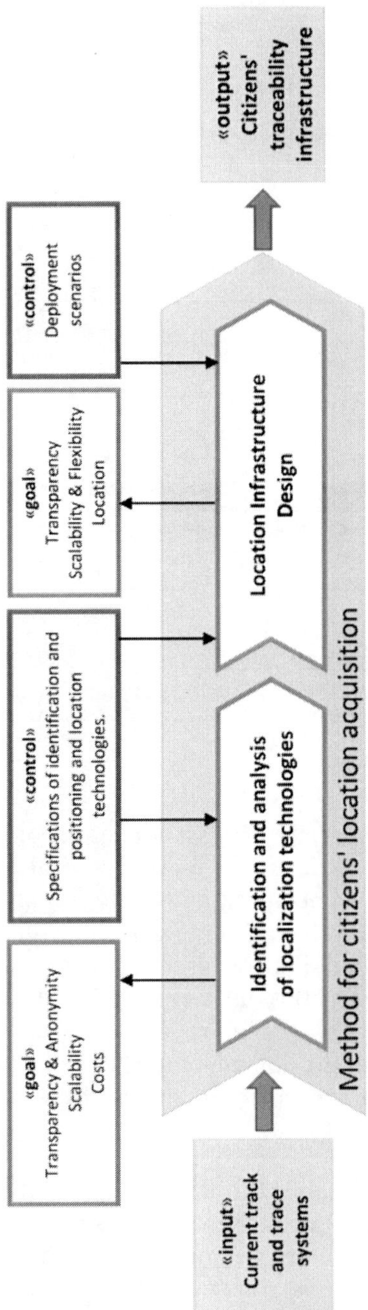

Figure 3. Modelling with Eriksson-Penker notation of Design of the method for citizens' localization acquisition process.

Table 2. Main features of communication technologies.

Technology	GPS	RFID/NFC	Wireless Networks
Main architecture	Triangulation positioning by satellite.	Set of antennas or readers and receivers.	Set of antennas or readers and receivers.
Communication	The user receiver obtains the signal from satellites and calculates the position.	Readers inspect receivers to determine whether they are present.	Receivers report that they are present.
Operating frequency	1100 MHz to 1600 MHz	Active: 455 MHz, 2.45 GHz, 5.8 GHz; Passive: 128 KHz, 13.6 MHz, 915 MHz, 2.45 GHz	Wi-Fi: 2.4 GHz, 5 GHz; WiMax: 2.3 GHz, 3.5 GHz; Cellular mobile: 800 MHz, 1900 MHz, others.
Cover	Worldwide; Outdoor environment	Depending on antenna network deployed. Outdoor and indoor environments	Depending on antenna network deployed. Outdoor and indoor environments.
Range	Worldwide	Active: ~100 m; Passive: 0 to few meters	Wi-Fi: 30 to 100 m WiMax: ~50 km; Cellular mobile: ~35 km
Power consumption	Very high	Passive tags receiver: Very low.	Very high
Deployment Costs	Satellites: Already deployed and free to use; Present in mobile devices	Need to deploy readers network. Present in mobile devices and other everyday user accessories.	Need to deploy antenna network. Present in mobile devices and other user accessories.
Localization	Receiver position.	Antenna position.	Antenna position.
Transparency and Anonymity	Low	High	Low
Usual application	Navigation, topography, land levelling, *etc.*	Identification, access control, payment, *etc.*	Internet access and communication services.

It is true that in terms of cost or ease of deployment of the infrastructure, there is already a high degree of integration of devices with GPS or Wi-Fi connectivity among the population. However, the energy cost of their use is high and its use may only for track and trace reasons not be justified. Finally, all the technologies are scalable to the deployment of infrastructure used, although the GPS positioning has an advantage since it does not require additional elements apart from the satellites. However, in the domain of the problem studied, the use of GPS is limited in indoor areas which should also be analyzed to manage their resources.

For all these reasons, among the analysed alternatives our proposal is to use RFID communication technology to track and trace citizens. As mentioned in the previous section, these technologies are having an expanding and continuous implementation and currently many devices incorporate it [78]. This allows the use for this purpose of the tag park that is becoming larger and it is currently already installed in many applications such as payment cards [63], clothing [64,65,66,67], access to transport services [40,62] or tourism [60].

In most previous applications the user's identity is not known and therefore it transmits only the identifier of the RFID tag worn by the user (in most cases, the tag could be attached to interchangeable everyday elements such as an umbrella or book which could be carried by different users). Consequently, the system doesn't know the correspondence between citizens and RFID tags. This technology does not allow a user to refrain from providing the location, since he or she will not be able to tear off the RFID tags from their clothes, accessories or cards. In that sense, using this technology for traceability should be treated similarly to CCTV which already has appropriate legislation: Deployed coverage infrastructure can be limited to specific areas, both indoors and outdoors and always inform the user about it, *etc.* However, there are also methods to preserve anonymity with RFID technology [53,79,80] which can be implemented in open city environments. In addition, the system may reduce the acquisition of unnecessary information which might involve intrusion into user privacy.

This technology provides installation options with passive or active receivers according to the desired range and complexity of the receiver. In the case of passive receivers, no power supply is required for transmitters [76,77], which reduces the drawbacks of consumption and portability for the user and enables its use for extended periods of time.

Regarding the economic cost of the proposal, the utilization of existing tags among the population reduces the cost of deployment. In this aspect, the physical nature of communication technology enables the use of metal urban furniture items as card reader antennas [38]. In this way, elements such as street lights, billboards, traffic signs or traffic lights can all operate as antennae. In addition, most of them are connected to the network, and then they can reduce the cost of the necessary elements for the information transmission towards the central system.

Finally, in terms of the ability for concurrent reading, the selected technology is able to work with a high level of traffic to simultaneously capture the location of a large number of tags located in the antenna reading radius. For example, the standard reader used in the experiments has a throughput up to 400 concurrent readings per second [81].

3.1.2. Location Infrastructure Design

The RFID technology mentioned in the previous section consists of two types of elements: The antennae or readers that emit the signals and receivers or tags. As mentioned, in the passive mode of this technology, receivers do not require power and can be embedded in objects familiar to users such as keychains, bracelets, keys or plastic cards worn by users.

Although this technology was conceived for identification processes, nowadays, it is also used for location tasks. Concerning this topic, research works are focused on the design of reader networks where each of them has associated geographical coordinates. Thus, from the received readings and deployed topology, it is possible to determine the position of the users (*location*).

The aim of this process is to design a network of RFID sensors deployed throughout the urban area of interest for reading the locations of citizens (e.g., shops, streets, avenues, parks, government buildings, *etc.*). The advantages associated with this technique make it easy to adapt to heterogeneous contexts (*flexibility*), and both indoor and outdoor environments (*transparency*). Moreover, adding new readers to the network is feasible according needs of the urban area management (*scalability*). The next figure (Figure 4) schematically shows the A-part of the overall architecture such a network of RFID sensors deployment in a city environment.

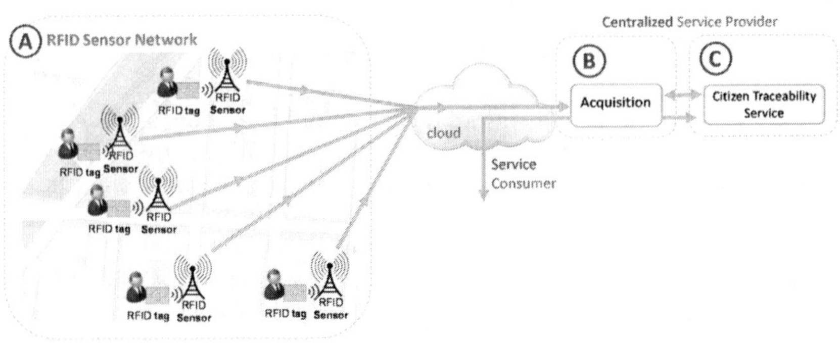

Figure 4. RFID sensor network deployment.

According to the features of RFID technology, it is possible to configure readers with coverage radii of several meters [76,77], so each of these readers can cover the sidewalk of a street entirely, as shown in Figure 5. The purpose is that collaboration of citizens for the activation of tags is not required and all they have to do is allow that they be worn on their clothes, carried in a pocket, bag or wallet in a transparent manner and interact as they move without the user being aware of it. Thereby, the tags worn by citizens are activated when they walk down the street and pass by one of the installed antennas.

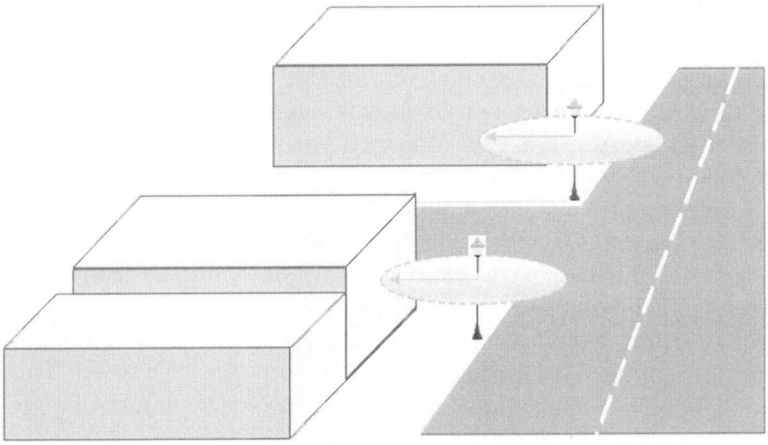

Figure 5. Installation and sensor coverage example.

The grid topology must be dense enough to cover the circulation in areas of interest. For practical reasons, tags used must be in some application that is already widespread and accepted by the citizens, while as mentioned, the new antennas can be complemented by already installed infrastructure.

With regards to the deployment of this additional infrastructure various criteria may be taken into account. For example, for practical criteria it may be decided to cover only areas of special interest in the city (downtown, tourist areas, *etc.*) or for economic criteria it may be decided to limit the grid to some areas of the city. In this regard, there may be options for installation of mobile antennae that can move from one part to another of the city when the area is sufficiently studied.

3.2. Method for Communication and Structuring of Citizens' Location

In the analyzed works based on RFID technology, the proposed solutions are focused on restricted areas, such as supermarkets, homes, hospitals, logistics centers, places of entertainment, *etc.* where the information transmission is performed directly through private intranets (not on public networks like the Internet) without reliability capabilities. In these proposals only an RFID tag type is used and processing needs have not been considered. However, after this analysis, it was identified that one of the challenges of the proposed system was the heterogeneity and lack of environmental control where it is mandatory to act in order to acquire the citizens' location information to build the movement flows. In this sense, any space in the city can be improved, without knowing the number or the type of tags to read or the capacity limits of the required processing. In this new scenario, it is necessary to propose a system with the technology and infrastructure indicated in the previous section, capable of distributing the logic of the system to ensure the workload processing, transmission and storage and structuring of information can be carried. This will establish the citizen flow and, therefore, the objectives will be addressed as outlined in the new scenario.

This is the objective of the task described in this section. More specifically, the goals which these process should cover are: To be capable of acquiring and processing the information independently of environment or data structure of the RFID tags (*transparency*), regardless of the number of readings performed by the park of sensors deployed in the network (*scalability*); to ensure the delivery and reception of each location to generate the different citizen flows (*reliability*) without citizen identification data (*privacy*); to ensure that the information is accessible from any place and on any device for the service consumers (*availability and versatility*); to offer complete information about the citizen movements, despite the lack of certain individual locations (*completeness*); to manage the systems resources in an optimal way (*efficiency*) and to distribute the load (*performance*), due to the high amount of location information that every reader will be able to receive and send to the central system; and, finally, to be interoperable in order to obtain information from different location formats or methods (*interoperability*), as the system is intended to take advantage of the already deployed elements and incorporate them for users who could might work with different protocols or formats.

In this sense, an analysis of techniques, patterns and design principles that guide and control the transformation process to achieve the proposed system and reach the mentioned goals has been necessary. The result of the process is a distributed architectural model based on the use of RFID Smart Sensors networks whose functionality will solve the problems (A-part of overall architecture), and a centralized cloud-based system which complements the information acquisition features. The RFID Smart Sensors basically consist of an RFID reader with computing and communication capabilities. The centralized cloud-based system will focus on the generation (B-part of overall architecture) and provisioning of citizens' flow (C-part of overall architecture) in an open and transparent way. The result of this process presents a reference model based on the integration of patterns and software design techniques. The novelty of this method lies in its combination and slight modifications to address the underlying problem and to solve existing deficiencies in current systems. This combination satisfies the use of RFID technology in uncontrolled and heterogeneous environments for determining the flow of citizens' movement in order to provide an efficient management of the resources of the cities. Figure 6 describes the processes involved in it, according to the methodology used with its main goals.

Next, we will describe the specifications of these elements in order to achieve the information acquisition, structuring, and communication goals.

3.2.1. Design of RFID Smart Sensor Network

Although there are related works which propose distributed architectural models based on sensors network to get the individual locations of the citizens (for example [42,45,48,50,53,56,58], these approaches do not cover necessary aspects (reliability, privacy, performance, efficiency, availability and transparency) to give a solution to the problem described. One of the main contributions and novelties of this paper is the provision of intelligence to sensors. This device is mainly responsible for reading the locations and ensuring the delivery of such information. So that, in order to provide the right route generation from each movement, it is necessary to design the infrastructure as a network of RFID Smart Sensors which include the necessary rationality to track citizens. Therefore, each Smart Sensor consists of an RFID reader, together with an element of embedded processing that provides computing and communication capabilities to support the necessary functionalities to reach established goals. These functionalities are supported by eight modules such as is shown in the schematic architecture of the RFID Smart Sensor (Figure 7).

One of the main functionalities included in the RFID Smart sensors is the capacity to receive information from different RFID tags types, transparently. About this issue, only the proposal presented in [45] proposes a solution that allows the capture of the location information from various types of RFID tags through the use of ontologies. Although this proposal would require a prior knowledge of the structure of information stored in the RFID tag to be read, an aspect that could mean that the management of ontology is untenable. It should be necessary to know all the existing tags types or every time is detected a new one to introduce the information into the ontology.

The remaining works analysed can only capture tag information provided by themselves. Since it is required in the problem proposed, our solution is not dependent on the RFID tag. In the Smart Sensor proposed, it has included functionality to obtain, transparently, full information as mentioned, from the generation of a code from the information that has been read to identify the location of form unique. When a new acquisition of the same tag arrives, the same code would allow traceability is created. Although a security system is included to increase the privacy of citizens that is every day changes the code generation. There is no wish to know if it is the same person carrying the tag, but the traceability of all citizens. These functionalities are performed by the *Location acquisition module* of the Smart Sensor described in Figure 7.

To perform reliability issue, we will implement the integration pattern known as guaranteed delivery [82]. This pattern determines an architectural design which prevents the loss of information messages transferred among the computer nodes that make up a distributed system. The pattern is based on the previous persistent message storage from a node in local disk. Until the destination node does not confirm message reception and storage, the source node will not delete the message from its local storage system. If the destination node never confirms message reception, the source node would retry sending it. Only with the use of the mentioned pattern, the complete delivery of the data cannot be ensured. The data could be sent and, because of the existence of communication problems, message reception confirmation may not arrive at the RFID station. In this case, if the message is not transmitted, the data would not be deleted and there would be a loss of information; or if the message were resent, data would be duplicated. To solve this issue, we focus on the correlation identifier pattern [82]. Each submission message will be marked with an identifier formed by a time stamp, which can establish the correlation for further processing, and by the identifier of the source RFID reader. Once communication is restored, the message would be sent again, being the acquisition module in the server the responsible for verifying information duplication. This way, we can ensure the complete reception of the necessary information to infer citizen flow patterns without losing or duplicating messages. This is another novelty functionality achieved through of patterns integration included in the RFID smart sensors. These functionalities and patterns are implemented by *persistent module, location message sender* and *update module* of the Smart Sensor shown in Figure 7.

Another key aspect to ensure delivery reliability in a distributed system, and it is not contemplated by the rest of the research works that provide RFID sensors distributed architectures, is the time synchronization of the system components. Most of the analysed proposals are focused on the tracking process [45,46,47,48,49,50,51,52,53,54,55,56,57,58]. The proposals focused on traceability process do not contemplate this question [42]. However, this aspect is necessary to achieve the tracing process. In order to generate reliable routes to manage resources in a better way, it is necessary to know the specific moment of the citizen location. A device provided with a badly synchronized clock would be sending the location at a wrong moment, the route information would not be valid, and it could even lead to wrong decisions. There are many algorithms that solve the problem of time synchronization, and choosing a particular one depends on the required accuracy of the system, the resources to be used, and the information that we want to obtain. For the present proposal, we will reject those systems that work with relative times and logical clocks, because, in our case, it is essential to synchronize all the elements of the system with a UTC reference system.

Depending on the type of network used and the precision required, we will have a great number of methods, among which are classical solutions such as Network Time Protocol (NTP), Simple Network Time Protocol (SNTP), and Time Transmit Protocol (TTP), or more advanced solutions such as Reference Broadcast Synchronization (RBS), Time Protocol on Sensor Networks (TPSN), and Flooding Time Synchronization Protocol (FTSP) [83,84].

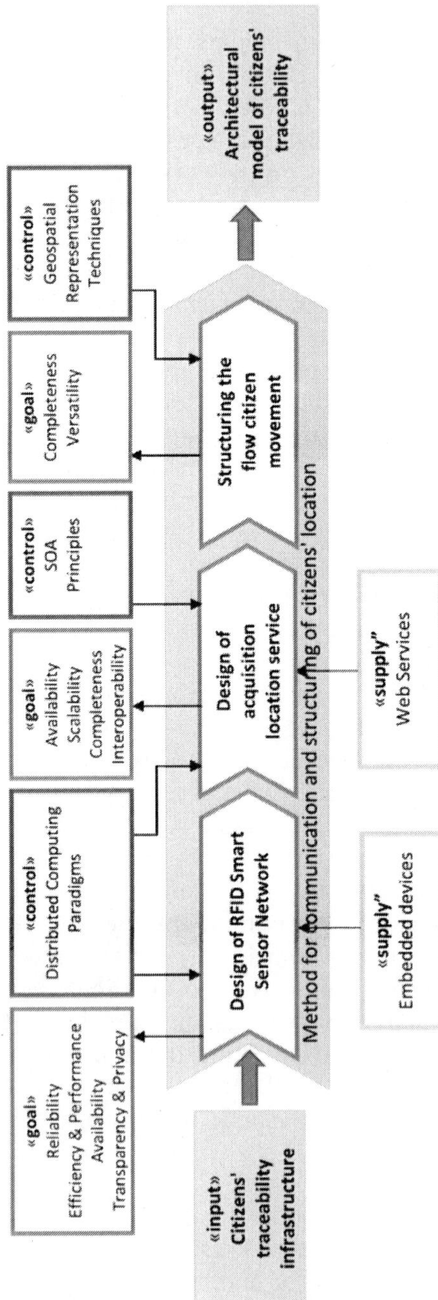

Figure 6. Modelling with Eriksson-Penker notation of the details of the process *Design of the method for communication and structuring of citizens' location.*

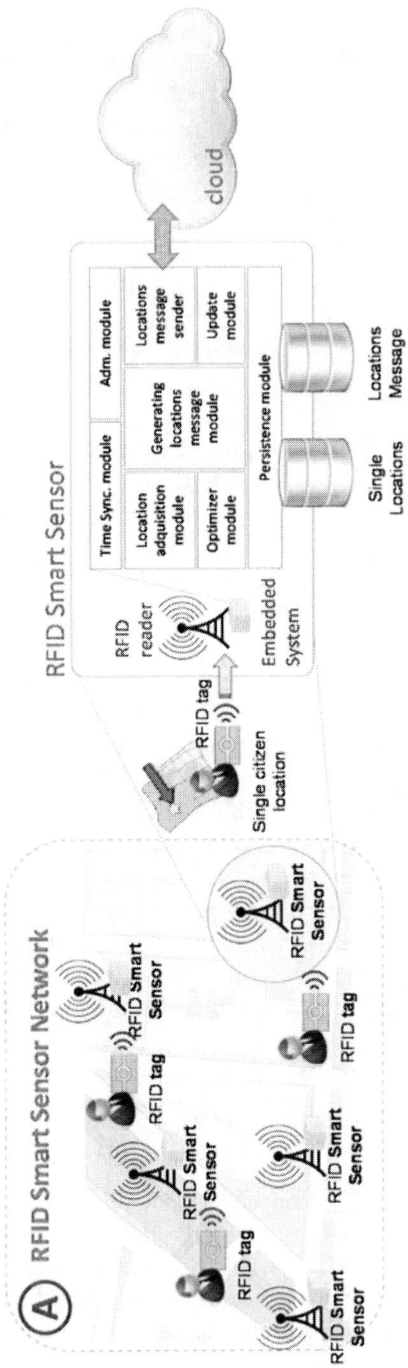

Figure 7. RFID smart sensor architecture.

In our case, the abovementioned patterns, together with time synchronization, ensure the reliability in the data delivery process. Also, the precision of the synchronization may perfectly tolerate a few-second lag, as the route would be relative to the local level and a range of several RFID readers would not appreciate such a lag. Moreover, once the server receives the location information from several neighbouring readers, there could be a data optimization and standardization regarding time. In the RFID Smart Sensor, this functionality is performed by the *time synchronization module*. This module is an autonomous and independent component that is executing periodically in background to update the clock system (Figure 7). In addition, the design proposed includes the following considerations in order to reduce the communication and management costs of the RFID smart sensor: Firstly, it is necessary to achieve an optimal transmission rate to take advantage of the bandwidth, which will be determined by the communication protocol established in the sensor. If we send the information via HTTP, the submitted location information could be less than the message wrapper of the communication protocol used. This circumstance would trigger penalties in each message sent. In our proposal, the number of locations in each sent message would be configurable in the Smart Sensor in order to achieve the best forwarding rate depending on the transport protocol used and the available bandwidth. The RFID Smart Sensor includes an *administration module* to manage and parameterized remotely the network sensors; secondly, the storage capacity of the Smart Sensors is limited, so it is essential to minimize local storage. In this sense, there would be a small modification of the guaranteed delivery pattern, in which not only would the message be erased to get the reception confirmation, but also the location data that were included in the message. With this design, once the RFID smart sensor obtains the location of a citizen, it stores it locally, creates a message from the citizen location data which is registered in the database, and stores them again. When the message is sent to the information system of citizen flow and is stored, the system returns reception confirmation to the client and the message is finally erased from the local system. This process is performed by the *update module* of the RFID smart sensor.

The remaining of the research works performs the communication process of the location data in controlled environments regardless of the communication reliability. For this reason, they do not propose the persistent capability at the sensor level and they have not contemplated the resources optimization of the sensors as our proposal.

Below, the proposed design works according to the following procedure: as the RFID reader is reading different citizen tags, these are received by the local location acquisition device through the acquisition module that performs the role of Coordinator.

Whenever an RFID tag content is received, the timestamp (only year/month/day) is attached to it and then the hash function is applied. The result of this action will be used to perform anonymity to the citizen. Then, this information is sent to the *optimizer module*, which is responsible for indicating if a citizen location must be stored or discarded depending if it is considered redundant data or not. If the information is valid, the *acquisition module* stores it in the local location database (*persistent module*) and informs the *generating location message module*, which uses the data to control if it reaches the optimal number of acquired locations to create a message that will be stored in the location message database. When the *location message sender* detects that there are new messages, it begins to transmit them by connecting to the *citizen flow generation service or acquisition location service*. The message will send the data through the available transport protocol.

At this moment, the RFID smart sensor will wait to the acknowledgement message location from *citizen flow generation service*. When the sensor receives it, then its update module is invoked, deleting the corresponding message and their individual locations from the database in the same transaction. This operation allows optimizing resources. If during a time previously configured the message acknowledgement was not received, the location message sender would return to try to send the message, as many times as was configured the device parameters. Before, the correct operation of network communications has to be checked.

3.2.2. Design of Acquisition Location Service

This service must provide an interface for data acquisition, which allows receiving location messages from users of smart sensors ensuring scalability and availability of the system, the reliability in data delivery and the interoperability with different types of sensors. This service corresponds with B-part of the overall architecture. The following figure shows the internal details of this module in conjunction with the smart sensor.

Scalability will be determined by two factors: firstly, the number of citizens who could interact with the system could grow indefinitely and, therefore, the data size to be acquired by the system could also grow in a reliable way. Secondly, the number of sensors deployed for the acquisition of citizen information may be high and could also grow considerably in order to obtain more data and to ensure accuracy in the flow determination. These two factors could make the citizen flow service provider become a bottleneck when it comes to receiving and processing the information. In order to avoid these situations, we propose two solutions that have been implemented: The first solution would be to separate the process of receiving and storing location messages, which requires less processing, from the transformation and processing of location data included in the messages. This solution would be provided by the message oriented middleware (MOM) paradigm or the software asynchronous queuing design pattern [85] (MOM component in Figure 8), establishing a point-to-point model based on message queues as an interface to the *location acquisition service*. This is a distributed computing model, which is asynchronous and transactional, and which acts as an intermediary message in the communication between a sender, the RFID Smart Sensor, and a consumer, the *citizen flow service provider*. In addition, this pattern enables the messages to remain, allowing us to implement the abovementioned guaranteed delivery pattern to ensure the reliability of the message communication; the second solution, which is complementary to the previous one, consists of the introduction of clustering techniques for load balancing using multiple nodes to be distributed and placing a messaging system for data acquisition in each of these nodes. Thus, the burden of data acquisition would be fairly distributed, and even the implementation of new acquisition nodes would be immediate.

Although the use of MOM paradigm has been proposed in [45] to provide scalability in indoor tracking system processing, in this proposal the communication reliability between sensors and central system and fault tolerance of the central system are not contemplated. Moreover, no aspect of high availability and load balancing is contemplated on this functionality as the needs that require scalability and processing are limited by the application environment.

In relation to the availability, a centralized system collapse or failure may involve the loss of location data and, therefore, reduces the validity of the generation of citizen flow patterns. To solve this problem, clustering techniques could be added to the acquisition module in order to achieve high availability. This way, if one of the nodes falls, another one from the cluster would replace it.

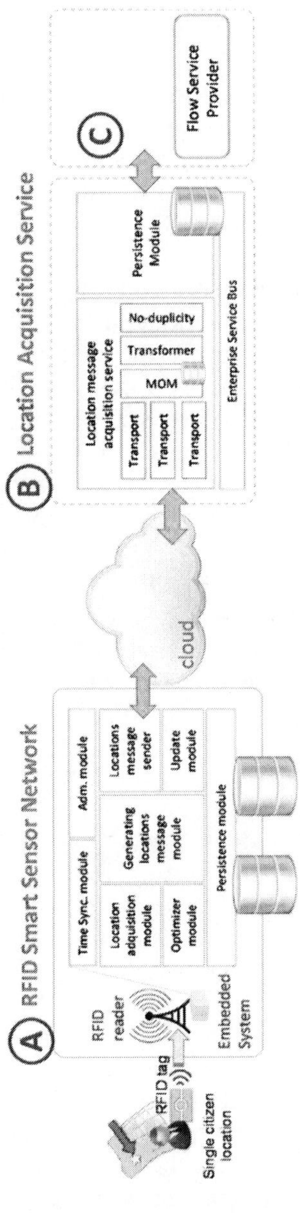

Figure 8. Location acquisition service architecture.

Finally, another important issue to be considered when location data are sent from RFID smart sensors is the heterogeneity of the existing client devices in the market and the communication protocols they use to transmit data. There are RFID readers which have computer systems that allow communication through standards which are known and widely extended. However, others only include serial communication and a small low-cost embedded device should be included to transmit information. In addition, although the proposal mainly relies on collecting and acquiring citizen locations through RFID technology, the system could consider acquiring data through other types of location and positioning systems such as GPS or legacy data systems that provide location information. In turn, this would lead a diversity of message formats and data models.

In those cases, we have considered creating an infrastructure based on protocol bridging patterns, data model transformation and data format transformation [85]. The protocol bridging design pattern is implemented by different transports modules (Figure 8), although, by default, the transport module of the system described in this section is based on Rest style following the SOA paradigm. The Rest service contract has been defined by means of *Rest APIs Modelling Language*, RAML (Figure 9). The other two patterns are implemented by the *transformer* component shown in Figure 8. The goal of the transformers is to unify the message in order to be the performance of citizen traceability service provider. In this way, we can dynamically solve the heterogeneity problems of the message structure, the data format and the communication protocol used for respectively sending the acquisition module.

When the message is received through the correct transport module, the message is stored in a persistent management system (point to point queue messaging model based on the pattern *asynchronous queuing* [85]). Messages are stored waiting to be consumed and processed. From this moment, the system sends an acknowledgement message to the RFID smart sensor to confirm location message reception.

It could happen that the message had been received by the system, but the acknowledgement response sent by the system fails because the communication was interrupted. In this case, a message previously processed could be processed another time. This is the function of the *no-duplicity* module, to filter the messages and to avoid their reprocessing using the message identifiers (message_id and rfid_reader_id). When a message is received from the queue of the message system (MOM), their structure and data format is checked. Depending on this information could be possible the use of transformers (*data model* and *format transformation* patterns) [85].

```
#%RAML 0.8
---
title: Citizens Location Service
baseUri: http://www.dtic.ua.es/v1.0
version: v1.0

/citizenlocations:
  post:
    description: create locations read from RFID smart sensors
    protocols: [HTTPS]
    body:
      application/json:
        example: |
          {
            "idrfid":1,
            "location":[
                 "lat":38.384993156837425,
                 "lng":-0.5133978999999727
            ],
            "locations":[
                 "citizen":["cid": 101010101,"pw": 64,"ts":1443723690],
                 "citizen":["cid": 101010101,"pw": 62,"ts":1443723695],
                 "citizen":["cid": 203330107,"pw": 62,"ts":1443723695]
            ]
          }
    responses:
      201:
        description: Locations have been successfully created.
        body:
          application/json:
            example: |
              {
                "messeage": "Locations have been successfully created."
              }
      400:
        description: Locations have not been created.
        body:
          application/json:
            example: |
              {
                "message": "Locations have not been created."
              }
```

Figure 9. RAML contract of location message acquisition service.

All modules of the location acquisition module are supported by *Enterprise Service Bus* (ESB) platform. ESB is a powerful infrastructure for software integration and communication that enables the interoperability between the described modules.

3.2.3. Structuring the Citizen Movement Flow

From the requirements analysis performed in the preceding paragraphs, in this process we must identify the information required to ensure the generation of reliable citizen flows which will allow making decisions in the management of the city resources. In order to do so, the main goals are the provision of complete information concerning the geographical area under study (*completeness*), with the possibility of segmenting the information depending on the time of day, the day of the week and the time of year (*versatility*).

The most basic and essential entity in order to determine the citizen flow of a citizen, will be the location (L). This entity will be defined by the following tuple:

$$L \equiv \{rfss\ (lo, la), cid, pw, ts\}$$

where:

- *rfss* (*lo*, *la*) represents the location of the RFID Smart Sensor determined by its longitude and latitude, which will allow us to know the position of the citizen on the map.

- *cid* represents the radio-frequency identification which is carried by the citizens and which allows us to identify them in a unique and anonymous way.

- *pw* it is the power of a certain card reading signal which will allow us to know the proximity to the RFID Smart Sensor antenna. Based on this, the information on the RFID Smart Sensor will be preprocessed in order to eliminate redundant locations with no added value which belong to the same user and will help determine possible directions of a citizen or trajectory changes with the data from other neighbouring RFID Smart sensor.

- *ts* it is the moment in time in which the RFID tag reading took place.

This data will be sent to the centralized system by the Smart Sensor, following the mechanisms described above to ensure delivery. The other data entity needed to determine the citizen flow is the deployed RFID smart sensor network itself. This information, together with the individual locations, will be completed to determine the route taken by a citizen. The sensor network entity will be determined by the following elements:

$RFSSN \equiv \{RFSS_1, RFSS_2,...., RFSS_n\}$

$RFSS_k \equiv \{id_k, lo_k, la_k, N_k\}$

$N_k \equiv \{RFSS_i...RFSS_j\} \subset RFSSN$

where:

- *RFSSN* represents the set of RFID Smart Sensors that make up the sensor network.

- $RFSS_k$ is a specific sensor defined by its unique identifier (id_k), its position, determined by the longitude (lo_k) and latitude (la_k), and the set of its neighbouring smart sensors (N_k), which in turn will be a subset of the network (note that the geographical-neighbours of a sensor could not match with its real neighbours due to insurmountable obstacles or habits of movement of citizens).

The data received by the deployed sensor network should be structured and completed in order to build the citizen movement routes in the geographic area under study. Therefore, knowing the neighbouring sensors allows addressing potential locations losses on the route followed by a citizen and helps define the interpolation and completeness strategies according to methods known to do so.

The combination of the data collected will be used to generate on-demand citizen flows for specific criteria, such as time interval or geographical area, indicating their density or the possibility to foresee citizen movement flows according to behaviour patterns based on historical movement data.

3.3. Method for Provisioning Citizen Flow

The orientation of this research to the study of public spaces in the city has led us to design methods to provide the information generated openly and as a value added service. In this section existing technological paradigms are used to enable easy consumption by third parties. The next figure (Figure 10) depicts schematically this service (C-part of overall architecture) in conjunction with the rest of the centralized architecture.

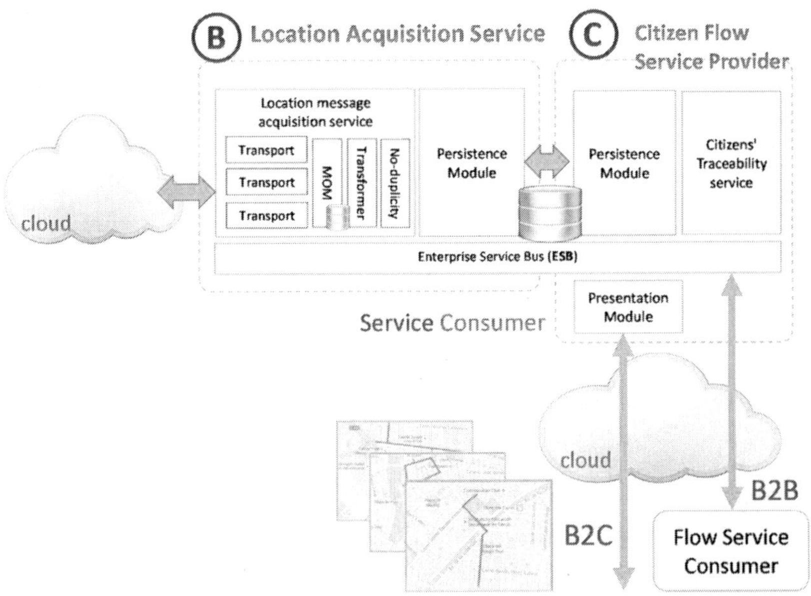

Figure 10. Location acquisition service architecture.

According to the foreseen, the goals focus, firstly, in the on-demand supply of the citizen movement flows based on the needs and search criteria for consumers (*segmentation*). Secondly, it is important that the value-added service can be used by external applications (*interoperability*). Therefore, this task has been divided in two stages: The design of data search filters and the supply of the system as an interoperable service.

In the first stage, several filters have been established to search based on the consumer needs. The first filter would be aimed at determining the action area where we want to make the decision. This action area could be defined by different indicators with a greater or lesser degree of accuracy such as a specific location, a specific location and a radius of action, a specific route, an area previously defined by the urban policy makers, or, finally, some irregular area determined by the service user. The second filter would determine the time period about which we want to obtain citizen flow data for the specific area and which would be given by parameters that represent different precision levels, from a short period of time in the same day to a long period in the long term, which could be months or years. Another filter would be based on the density of routes, allowing the information about specific routes that exceed or are below a threshold and that can contribute to the modification of the use of city resources. For example, a route with low density for a long period of time may lead to the modification of a public transport route.

As mentioned before, the second stage focuses on securing the interoperability of the service with the consumers. In order to do so, this service should offer the following two models: A first approach based on Business to Business (B2B) paradigm will allow connecting the consumer system with the service to incorporate it as another one of its organizational processes, but improving the current system by adding more decision data. Both cases will be based on the service supply according to the principles of the Service Oriented Architecture (SOA) paradigm, [85] implementing through Rest service technology and offering the resulting information using a format that is structured as JSON, so that it is easily treated. The service contract, modelling with RAML, is shown below (Figure 11). As can be seen, the service contract includes the filters defined previously to carry out searches by means of next parameters: Action zone, time period and flow density; a second approach, based on the Business to Customer (B2C) paradigm, oriented to offer a human-machine interface which will allow a user to interact directly with the system, for example through a thin client such as a Web browser. The access interface of this module defines the search filters described previously. When the Rest service receives these parameters, this one returns the citizens flows and their density drawn on cartographic map. In both cases, the citizen' flows will be generated from the information collected by RFID Smart Sensors and the information of the sensors topology grid dynamically and on demand, based on the input parameters representing the filters entered by the consumer. The generation of flows combined through individual routes of the same or multiple users may be caused by adding readings for the specific filters through a clustering process.

```
#%RAML 0.8
---
title: Citizen Flow Service
baseUri: http://www.dtic.ua.es/v1.0
version: v1.0

/citizensflows:
  uriParameters:
    zone:
    timeInterval:
    density:
  get:
    description: get citizens flows
    protocols: [HTTPS]
    responses:
      200:
        description: Get citizens flows.
        body:
          application/json:
            example: |
              {
                "flows":[
                     "flow":[
                        "density":20,
                        "color":"red",
                        "locations":[
                          "location":["lat":38.384993156837425, "lng":-0.5133978999999727],
                          "location":["lat":38.3850402521112, "lng":-0.5150179542449673],
                          "location":["lat":38.38557848104228, "lng":-0.5152110732940396]
                        ],
                     "flow":[
                        "density":10,
                        "color":"blue",
                        "locations":[
                          "location":["lat":38.38687357799348, "lng":-0.5135802902129849],
                          "location":["lat":38.3855120264542, "lng":-0.5130653060821255]
                        ]
                     ]
                ]
```

Figure 11. Citizen flow service RAML contract.

The calculation of these added routes can be performed through several procedures by applying a similarity function to them. For example, some of the functions of this type used are the comparison of the common points of the sequence, and the comparison of distances or times used in them. These procedures are beyond the scope that defines this architecture, although they are being considered in the line of work developed by the research group.

4. IMPLEMENTATION AND VALIDATION

4.1. Implementation

In order to be able to validate the proposal, a prototype of the architectural model shown previously has been implemented. The RFID Smart Sensor prototype is shown in Figure 12. The sensor used is marketed by *RFID Controls Enterprise* (Murcia, Spain). This is a highly configurable and flexible sensor with two antennas [81]. It has a maximum reading rate of 400 tags/s and its maximum reading potency is 1 W. This device even allows connection to two antennas through two 50 Ohm MCX connectors with 6.5 dB gain, providing circular polarization and a radiation pattern with 60°/60° beamwidth (Figure 12a).

(a) (b)

Figure 12. RFID Smart Sensor prototype: (**a**) RFID Smart Sensor with antennas; (**b**) RFID reader with an embedded system.

In order to provide the sensor with the necessary intelligence for implementing the aforementioned modules, an embedded device based on an ARM processor has been used. This device has a low energy consumption and it has been integrated with the RFID reader through its Ethernet interface. Specifically, the Cubieboard3 cortex-A7 dual-core 2 Gb embedded device has been used, provided by *Cubietruck Enterprises* [86] (Figure 12).

The RFID Smart Sensor modules designed in the architectural model have been implemented using C language. Reading results are saved in

CSV files and stored in a system directory. To implement the clock synchronization module the NTP protocol has been used, because is a reliable protocol with suitable error rates for our system. For this reason, an NFS server has been enabled, which is used to update the system clock by means of the synchronization module implemented in C language.

The proposed architecture of the citizen track and trace system is based on *Enterprise Service Bus integration infrastructure*. This solution is highly scalable and the software design patterns specified in the architectural model have been implemented over this platform. Specifically, the open source Mule ESB has been used, implemented two Mule ESB services flows to implement the *Location Message Acquisition Service* and the *Flow Generation Service*, shown in Figure 13a,b, respectively.

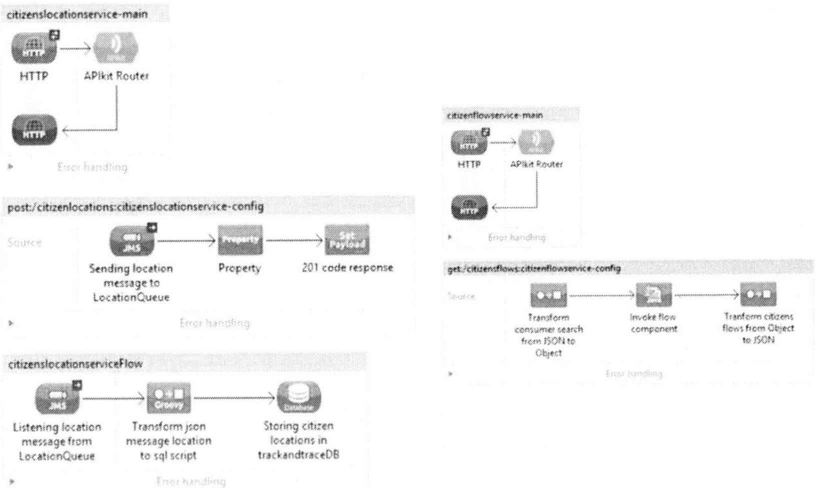

(a) (b)

Figure 13. Mule flow. (**a**) Location message acquisition service; (**b**) Citizens flows generation service.

In order to provide data persistence, the *MySQL community version* database manager has been used. The *Search Module* for flow generation has been implemented using Java Language. This module is embedded and invokes the Mule service flow. Finally, the *B2C*

Presentation Module has been developed through Web technologies like CSS, HTML5 and JavaScript using the AngularJS framework, in order to provide more accessibility and usability to the final users.

In order to achieve better analysis and interpretation of the experimentation results obtained in the case study described below, we have previously made a technical testing of the smart sensor and the RFID technology. The aim is to know beforehand their capabilities and limitations to certain situations that may occur.

First, we proved that with the power supplied by the sensor (1 W) and with the well-focused antenna the reader is capable of reading tags up to 10 meters away. Secondly, reading tests were performed simulating actual portability by citizens, *i.e.*, placing tags hidden in clothes, pockets, purses, handbags or backpacks. These experiments significantly reduce the sensor range down to 3 m, but a good reading rate is still maintained in these cases.

The most serious problem encountered for our purpose is that the human body is opaque to the RF signal, so that the tag will be hidden from the sensor when there is a human obstacle. This case can be common in urban settings with a high density of pedestrians. For example, in Figure 14a, the two persons on the left are in a shadow area produced by another person. However, this problem can be solved by distributing the antennas of the smart sensor in a configuration that minimizes occlusions and signal shadow areas. Following the previous example, Figure 14b shows a configuration that reduces the space not visible to the sensor.

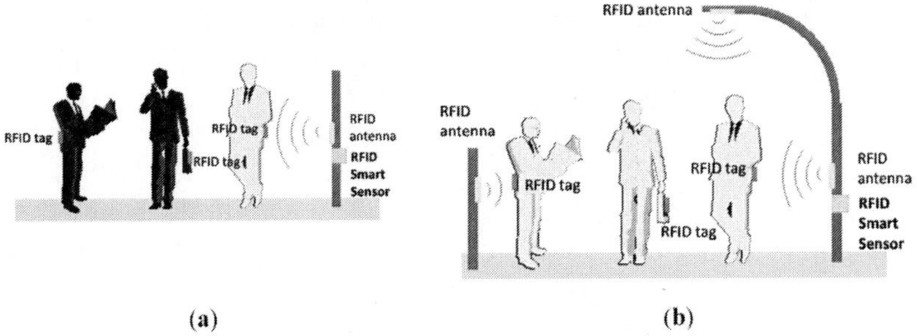

Figure 14. Configuration of the antennas and coverage of RFID sensors: (**a**) One antenna on the right side of a sidewalk; (**b**) Three antennas on the right, left and top of a sidewalk.

Regarding the other capabilities of the Smart sensor, the experiments proved the ability to synchronize with other sensors, the combination of multiple readings from the same person and the capability of sending information without loss. Through log systems of the proposal of the traceability of reading, we analyse sent and received citizens' locations and we verify that all locations readings are stored in the *RFID Smart Sensors Network* and they were received by *Location Acquisition Service*. In this way, the reliability of the proposed architecture is validated.

4.2. Case Study

In order to validate our proposal as well as to study a practical implementation of the concepts treated, a simulated use case in a real environment for tracking and tracing is proposed in this subsection. In our example, we consider the campus of the University of Alicante as a test scenario for the deployment of the proposed system. The case study will provide a better understanding of how various campus resources are used based on student and staff spatial movement and temporal behaviour. The ultimate goal of this case is to help the university improve management of the support services on campus.

The experiments performed will also result in a number of spatial and temporal visualizations of big data from these sources. The reason to carry out a project on a campus is the advantage of disposing of something similar to a little city, but more accessible. In a way, since it has different areas (rooms, labs, small shops, restaurants, open areas, *etc.*), diverse activities (academic, administrative, social, small health centres, sports...) the university campus can be seen as a small city. In addition, it has the facilities to make it as a perfect candidate to test our system under diverse conditions in order to find out citizen movement flow patterns. It is always far less complicated to locate antennas, RFID readers and the rest of technology within the university campus than throughout a city. In addition, it is possible to monitor the different states of every single location point, allowing testing of what are the best and most suitable places for the whole system. Therefore, in the matter of fact, we can see our system implemented in the framework of the campus as a prior step for a smart city which is much bigger, although it can be benefit from this previous stage.

The criterion for placement of the readers has been to monitor the most likely movements of the students (cafeteria-classrooms, classrooms-library, bus station-classrooms, *etc.*) as a preliminary step to a more dense deployment. Students were not informed that they were taking part in this experiment, only academic campus authorities were informed to request the appropriate permissions for the deployment of the readers. Due to this, no tags were distributed among individuals, but our intention was to first validate the transparency and anonymity of the readings, the ability to read tags and carried by people (on clothing, tickets, identification or transport cards, other RFID tags) without their cooperation and the capacities of transmission and processing of the architecture.

The experiment was conducted during one week. Through a visual monitoring of the antennas the number of students who passed through the sensor radius was counted. It was about 2K per day, with a tag read rate of about 5%. The amount of simultaneous readings of this data remained low, although, in most cases, when the tags were read by an antenna, they were read again by the next antenna on the same route.

From the study of these results, the reasons that lead the low rate of readings obtained can be: (1) the still low level of integration of RFID tags on clothing and other objects carried by users; (2) the lack of power in the readers to reach tags hidden in the wallet in the pocket of the user; and (3) the incompatibility of technologies with short-range tags (as NFC) to access ID cards or credit cards.

As to the first drawback (1), we think that will be resolved eventually, as the steadily increasing pace of implementation of RFID technology in everyday objects such as clothes or accessories will increase the reading rate. The second problem (2) can be overcome using directed antennas with a low angle of beamwidth coverage and high gain antennas rather than general purpose ones since in outdoor spaces, the signal can be dispersed more or noise sources can even appear [87,88]. In this regard, it is necessary to check regional regulations to find out the maximum allowable radiated power and antenna beamwidth in the UHF RFID bands. Finally, regarding the third cause (3), we think that this kind of RF technology belongs to a set of applications that should not be accessible, so to remain inaccessible, reinforcing anonymity of identity and user data.

In all cases, the readings collected and sent to the server correspond to a tag identifier on clothing or an accessory. This information consists, among others, of the following fields: An ID tag of different length, IP address of the reader, received signal power and DateTime (*i.e.*, 30396062c3c324800009e6 77, 169.254.247.184, −74, 03/02/2015 08:55:59.121). Under no circumstances is personal user information collected. Due to the functionality included in the *Location Acquisition Module*which codifies RFID tags content through hash functions, we could not get any data of those RFID tags. In addition, any of the read RFID tags could be related themselves between readings obtained from different days.

The central system is hosted on a server on the campus intranet. The network traffic generated and the system processing times are trivial for a standard workstation, as the number of simultaneous readings remained low all the time, reaching traffic peaks of up to 3 reading/s.

To view the results we used a web-based user interface, which shows the positions of the readers and users-route through a third party application (Google Maps JavaScript API v3). For this case study, the next figure (Figure 15) shows a map of the University of Alicante showing the installation locations of the reader antennas and two temporal visualizations along the day where one can distinguish the main flow pattern per hour.

The results show that the movement flows of students change throughout the day. In Figure 15a it can be seen that the antenna located near the tram and bus platform receives students first, and from there, they are read by sensors located in other areas of the university as they move around the campus, both via inbound and outbound scenarios. In these hours of the morning, the movements are primarily targeted towards classroom buildings, offices and the library. Figure 15b shows the movements at other times of the day, when students walk to the campus exit and to restaurants and cafes to have lunch.

In this case study, the students' movements read by the system may seem obvious at first glance, since they coincide with the normal movement at that time of day. However, this case study provides the starting point for further deployment in which the movements of students and staff of the university can be analyzed in more detail. This information can help managers to provide better services and improve user satisfaction by providing resources where they are needed at any given moment.

In light of these results, we can assert that this experiment verifies the validity of the proposed hypothesis regarding the capabilities of RFID sensing technologies to read the positions of citizens outdoors without needing any kind of collaboration. The correction of distributed architecture to collect, synchronize and to process readings received while maintaining the anonymity of the users involved, it is also validated. The workload has not been a problem in this context due to the low rate of simultaneous readings. It is expected that a larger-scale deployment will allow us to validate the system in this regard. This result provides a promising perspective in the development of IoT for interaction with users.

Finally, a summary of the main objectives achieved by the proposed architecture in comparison with the capabilities of the other proposals is shown in Table 3. The comparison is focused on the results obtained in the experimentation, indicating what objectives would be achieved with each proposal. The table illustrates that related works offer partial solutions to complete all the stated objectives.

5. CONCLUSIONS

The design of smart cities uses the information generated in the city itself for decision making aimed at improving the quality of life of citizens. Infrastructure planning is one of the most pressing challenges of the development of cities and usually the handling of this issue does not only correspond to practical criteria, but also social, political, historical or other aspects. Knowledge of citizens' flow movements is very valuable information for designing transportation routes, as well as the optimal positioning of resources. With this knowledge, it is feasible to perform infrastructure designs which can minimize deviations from real movement flows. It is also possible to increase transport dynamic strategies in accordance with these needs to meet the movement habits of people by time of day, day of week or time of year to increase the use of infrastructure and thus improve the comfort of the citizens. The latest ICT advances can offer promising solutions to this specific problem by providing efficient and effective ways for handling large amounts of unstandardized and distributed information.

The proposal described in this paper is the design of an architecture that can read, process and provide as a service the movements of individuals in a given geographic area based on real-time information acquired from the individuals themselves as they move along the city in an unassisted way to improve the resources management of cities.

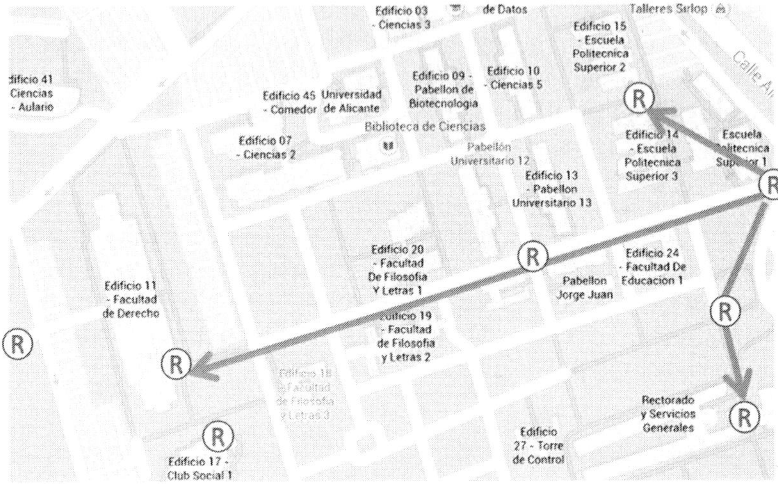

(a)

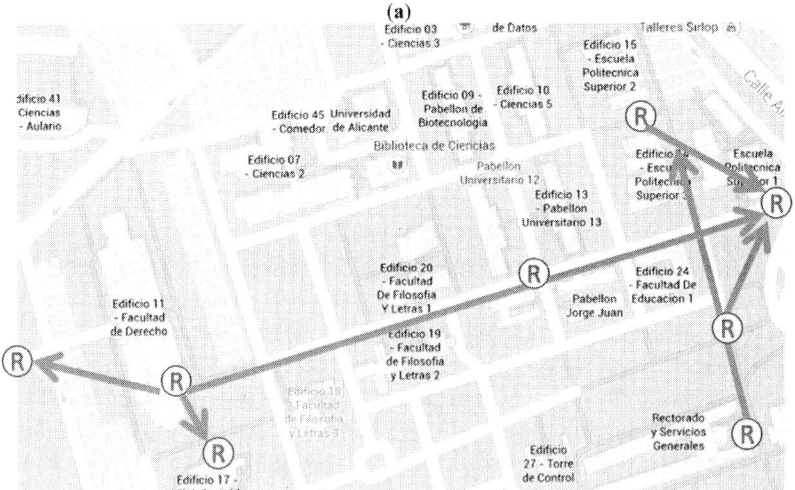

(b)

Figure 15. Temporal visualizations on the University Alicante campus with flow patterns: (**a**) temporal flow between 7.00 and 9.00 h; (**b**) temporal flow between 12.00 and 14.00 h

The research presented in this paper includes an exhaustive analysis of the state of the art of techniques for solving the citizen traceability problem in cities. Its conclusions have allowed us to identify limitations of the current methods and technologies on the basis of the following main criteria: Heterogeneity, maintaining anonymity, no user intervention, uncontrol and cost. A comprehensive architectural model has been conceived using RFID wireless technologies with passive tags present in many devices, accessories or the clothes of the citizens to perform track and tracing tasks in city environments. As a novelty, we have designed an RFID smart sensor to read, process and to send the information of the users' locations in a scalable and reliable way to ensure the delivery and reception of each location. Finally, we have developed a computational architecture which integrates all the elements and modules implied to provide cloud services of citizens' traceability for B2B and B2C, consumers based on SOA principles.

Table 3. Comparison of the obtained results between our proposal and the related works.

Research Work	Main Functionality	Transparency & Anonymity	Working Environments	Reliability	Energy Consumption	Scalability
Proposed architecture	Track & trace	Yes. No user interaction & Independency of RFID tags	Indoor & outdoor	Yes	No (passive tags)	Yes. (Distributed and decoupled approach)
GPS proposals [27–30]	Track & trace	No. User interaction	Outdoor	No	High	No
Wi-Fi proposals [16,34]	Track	No. User interaction	Indoor & outdoor	No	High	No
CCTV	Surveillance	No. Surveillance cameras	Indoor & outdoor	No	High	Yes
Science museum [42]	Track & trace	No. User/Object interaction & Dependency of own RFID tags	Indoor	No	No (passive tags)	No
Oyster Card [40]	Track & trace	No. Dependency of own RFID tags	Indoor & outdoor	No	No (passive tags)	No
Blind/inside RFID [47,53]	Track	No. User interaction & Dependency of own RFID tags	Indoor & outdoor	No	No (passive tags)	No
REACT [58]	Track	No. User/Object interaction & Dependency of own RFID tags	Outdoor	No	No (passive tags)	Yes
LANDMARC [41,46]	Track	No. User/Object interaction & Dependency of own RFID tags	Indoor	No	Yes (active tags)	No
Peer-to-Peer Networks [57]	Track	No. User/Object interaction & Dependency of own RFID tags	Indoor	No	Yes (active tags)	Yes
iWalker [44]	Track	No. User/Object interaction & Dependency of own RFID tags	Indoor	No	No (passive tags)	No
Other [45,48,50–52,54–56]	Track	No. User/Object interaction & Dependency of own RFID tags	Indoor	No	No (passive tags)	In controlled env. only [45,55]

The way of achieving of goals established and solution of problems raised have been the main contributions of this research. The solution has been in most cases to identify techniques and patterns that partially solve these issues to integrate them and offer a global solution. It is worth mentioning that the integration has brought changes in patterns and adjusting these elements to offer a comprehensive solution we have not found in the literature reviewed. In addition, these solutions have been encapsulated in functions and distributed in elements building the architecture to ensure the requirements of the problem. As a result, a network of intelligent sensors is provided, which is a novel solution for the current track and trace systems, fully coordinated and a central system is added to provide the citizens' flow.

The work involved in this research fits into the concept of Smart City aimed at improving lives of citizens using current technology of communications, cloud computing and smart sensing. In this regard, some challenges for future work arise, such as combining other communication technologies (GPS, Wi-Fi, *etc.*), integrating other sensing devices to improve monitoring (smart phones, *etc.*), studying mechanisms to optimize the network, new sensors deployed to cover the largest area possible or applying the presented architecture in a real city scenario. Moreover, the combination of these communication technologies with social media data such as Twitter and Facebook, will help develop broader patterns of citizen behaviour and compromise, service and facility utilization.

ACKNOWLEDGMENTS

We would like to express our acknowledgement to the *RFID Controls Company* for proving the necessary infrastructure to make possible the experimentation.

AUTHOR CONTRIBUTIONS

All authors were involved in the foundation items. All authors wrote the paper and read and approved the final manuscript.

CONFLICTS OF INTEREST

The authors declare no conflict of interest.

REFERENCES

1. Neirotti, P.; de Marco, A.; Cagliano, A.C.; Mangano, G.; Scorrano, F. Current trends in Smart City initiatives: Some stylised facts. Cities 2014, 38, 25–36.

2. Smart Cities Council. Available online: http://smartcities council.com (accessed on 9 June 2015).

3. OECD, Governing the City. Available online: http:// dx. dir. org/10.1787/9889264226500-en (accessed on 9 June 2015).

4. European Smart Cities. Available online: http://www.smart-cities.eu/ (accessed on 9 June 2015).

5. Fietkiewicz, K.; Pyka, S. Development of informational cities in Japan: A regional comparison. Int. J. Knowl. Soc. Res.2014, 5.

6. Fazio, M.; Puliafito, A.; Villari, M. IoT4S: A new architecture to exploit sensing capabilities in smart cities. Int. J. Web Grid Serv. 2014, 10, 114–138.

7. Atzoria, L.; Ierab, A.; Morabito, G. The Internet of things: A survey. Comput. Netw. 2010, 54, 2787–2805.

8. Mitton, N.; Papavassiliou, S.; Puliafito, A.; Trivedi, K.S. Combining Cloud and sensors in a smart city environment.EURASIP J. Wirel. Commun. Netw. 2012, 2012, 247.

9. Stankovic, J.A. Research directions for the internet of things. IEEE Internet Things J. 2014, 1, 3–9.

10. Gubbia, J.; Buyyab, R.; Marusica, S.; Palaniswamia, M. Internet of things (IoT): A vision, architectural elements, and future directions. Future Gener. Comput. Syst. 2013, 29, 1645–1660.

11. Choi, S.H.; Yang, B.; Cheung, H.H.; Yang, Y.X. RFID tag data processing in manufacturing for track-and-trace anti-counterfeiting. Comput. Ind. 2015, 68, 148–161.

12. Shia, J.; Lia, Y.; Heb, W.; Sima, D. SecTTS: A secure track & trace system for RFID-enabled supply chains. Comput. Ind.2012, 63, 574–585.

13. Kang, Y.S.; Lee, Y.H. Development of generic RFID traceability services. Comput. Ind. 2013, 64, 609–623.

14. Muñoz-Gea, J.P.; Malgosa-Sanahuja, J.; Manzanares-Lopez, P.; Sanchez-Aarnoutse, J.C. Implementation of traceability using a distributed RFID-based mechanism. Comput. Ind. 2010, 61, 480–496.

15. Todorovica, V.; Neagb, M.; Lazarevica, M. On the usage of RFID tags for tracking and monitoring of shipped perishable goods. Proced. Eng. 2014, 69, 1345–1349.

16. Isaak, J. Cell Phone WiFi Used to Track Your Location. Available online: http://ieeessit.org/2014/01/16/cell-phone-wifi-used-to-track-your-location-2/ (accessed on 9 June 2015).

17. International Telecommunication Union. Aggregate Data 2006–2013 ICT Data for the World, by Geographic Regions and by Level of Development; ITU: Geneva, Switzerland, 2014.

18. Kelly, J.; Jones, P.; Barta, F.; Hossinger, R.; Witte, A.; Wolf, A.C. Successful Transport Decision-Making: A Project Management and Stakeholder Engagement Handbook; European Commission: Brussels, Belgium, 2004.

19. Allport, R.; Brown, R.; Glaister, S.; Travers, T. Success and Failure in Urban Transport Infrastructure Projects; KPMG Intternational: Amsterdam, The Netherlands, 2008.

20. Bickerstaff, K.; Tolley, R.; Walker, G. Transport planning and participation: The rhetoric and realities of public involvement. J. Transp. Geogr. 2002, 10, 61–73.

21. Galland, S.; Lamotte, O.; Gaud, N. MetroB: Evaluation and simulation of public transport system. In Proceedings of the IET International Conference on Smart and Sustainable City, Shanghai, China, 6–8 July 2011.

22. Giannotti, F.; Nanni, M.; Pedreschi, D.; Pinelli, F.; Renso, C.; Rinzivillo, S.; Trasarti, R. Mobility data mining: Discovering movement patterns from trajectory data. In Proceedings of the Second International Workshop on Computational Transportation Science, San Jose, CA, USA, 2 November 2010; pp. 7–10.

23. Batty, M.; Axhausen, K.; Giannotti, F.; Pozdnoukhov, A.; Bazzani, A.; Wachowicz, M.; Ouzounis, G.; Portugali, Y. Smart cities of the future. Eur. Phys. J. Spec. Top. 2012, 214, 481–518.

24. Schmitt, G. A planning environment for the design of future cities. In Digital Urban Modelling and Simulation; Springer: Berlin, Germany, 2012; pp. 3–16.

25. Bajaj, R.; Ranaweera, S.L.; Agrawal, D.P. GPS: Location-tracking technology. Computer 2002, 35, 92–94.

26. Moloo, R.K.; Digumber, V.K. Low-cost mobile GPS tracking solution. In Proceedings of the 2011 International Conference on Business Computing and Global Informatization, Shanghai, China, 29–31 July 2011; pp. 516–519.

27. Trasarti, R.; Rinzivillo, S.; Pinelli, F.; Nanni, M.; Monreale, A.; Renso, C.; Pedreschi, D.; Giannotti, F. Exploring Real Mobility Data with M-atlas. In Machine Learning and Knowledge Discovery in Databases; Springer: Berlin, Germany, 2010; pp. 624–627.

28. Yuan, J.; Zheng, Y.; Zhang, C.; Xie, W.; Xie, X.; Sun, G.; Huang, Y. T-drive: Driving directions based on taxi trajectories. In Proceedings of the 18th ACM SIGSPATIAL International Conference on Advances in Geographic Information Systems, San Jose, CA, USA, 2–5 November 2010; pp. 99–108.

29. Trasarti, R.; Nanni, M.; Giannotti, F.; Pedreschi, D.; Renso, C. A query language for mobility data mining. Int. J. Data Warehous. Min. 2011, 7, 24–45.

30. Iqbal, M.U.; Lim, S. Privacy implications of automated GPS tracking and profiling. IEEE Technol. Soc. Mag. 2010, 29, 39–46.

31. Michaela, K.; Clarke, R. Location and tracking of mobile devices: Überveillance stalks the streets. Comput. Law Secur. Rev. 2013, 29, 216–228.

32. Clarke, R.; Wigan, M.R. You are where you've been: The privacy implications of location and tracking technologies. J. Locat. Based Serv. 2011, 5, 138–155.

33. Dahunsi, F.; Dwolatzky, B. An empirical investigation of the accuracy of location-based services in South Africa. J. Locat. Based Serv. 2012, 6, 22–34.

34. Zhang, Y.; Li, L.; Zhang, Y. Research and design of location tracking system used in underground mine based on WiFi technology. In Proceedings of the International Forum on Computer Science-Technology and Applications, Chongqing, China, 25–27 December 2009.

35. Chen, P.-C. A cellular based mobile location tracking system. In Proceedings of the IEEE 49th Vehicular Technology Conference, Houston, TX, USA, 16–20 May 1999.

36. Mitra, S.; DasBit, S. On location tracking and load balancing in cellular mobile environment-a probabilistic approach. In Proceedings of the International Conference on Electrical and Computer Engineering, Dhaka, Bangladesh, 20–22 December 2008; pp. 121–126.

37. Chawla, V.; Ha, D.S. An overview of passive RFID. IEEE Commun. Mag. 2007, 45, 11–17.

38. Weinstein, R. RFID: A technical overview and its application to the enterprise. IT Prof. 2005, 7, 27–33.

39. Ni, L.M.; Zhang, D.; Souryal, M.R. RFID-based localization and tracking technologies. IEEE Wirel. Commun. 2011, 18, 45–51.

40. Bagchi, M.; White, P.R. The potential of public transport smart card data. Transp. Policy 2005, 12, 464–474.

41. Ni, L.M.; Liu, Y.; Lau, Y.C.; Patil, A.P. LANDMARC: Indoor location sensing using active RFID. Wirel. Netw. 2004, 10, 701–710.

42. Kanda, T.; Shiomi, M.; Perrin, L.; Nomura, T.; Ishiguro, H.; Hagita, N. Analysis of people trajectories with ubiquitous sensors in a science museum. In Proceedings of the IEEE International Conference on Robotics and Automation, Roma, Italy, 10–14 April 2007; pp. 4846–4853.

43. Pandit, A.A.; Talreja, J.; Mundra, A.K. RFID tracking system for vehicles (RTSV). In Proceedings of the First International Conference on Computational Intelligence, Communication Systems and Networks, Indore, India, 23–25 July 2009.

44. Kulyukin, V.; Kutiyanawala, A.; LoPresti, E.; Matthews, J.; Simpson, R. iWalker: Toward a rollator-mounted wayfinding system for the elderly. In Proceedings of the IEEE International Conference on RFID, Las Vegas, NV, USA, 16–17 April 2008.

45. Li, Z.; Chu, C.-H.; Yao, W. SIP-RLTS: An RFID location tracking system based on SIP. In Proceedings of the IEEE International Conference on RFID, Las Vegas, NV, USA, 16–17 April 2008.

46. Han, K.; Cho, S.-H. Advanced LANDMARC with adaptive knearest algorithm for RFID location system. In Proceedings of the International Conference on Network Infrastructure and Digital Content, Beijing, China, 24–26 September 2010; pp. 595–598.

47. Willis, S.; Helal, S. A Passive RFID Information Grid for Location and Proximity Sensing for the Blind User; Technical Report Number TR04-009; University of Florida: Gainesville, FL, USA, 2004.

48. Iturralde, D. A new system based on web services and RFID for tracking people in a pervasive mining environment. In Proceedings of the Latin-America Conference on Communications (LATINCOM), Santiago, Chile, 24–26 November 2013.

49. Buccafurri, F.; Lax, G.; Nicolazzo, S.; Nocera, A. A privacy-preserving solution for tracking people in critical environments. In Proceedings of the 38th International Computer Software and Applications Conference Workshops (COMPSACW), Vasteras, Sweden, 21–25 July 2014; pp. 146–151.

50. Nohara, K.; Tajika, T.; Shiomi, M.; Kanda, T.; Ishiguro, H.; Hagita, N. Integrating passive RFID tag and person tracking for social interaction in daily life. In Proceedings of the IEEE 17th International Symposium on Robot and Human Interactive Communication, Munich, Geramny, 1–3 August 2008; pp. 545–552.

51. Germa, T.; Tajika, T.; Shiomi, M.; Kanda, T.; Ishiguro, H.; Hagita, N. Vision and RFID data fusion for tracking people in crowds by a mobile robot. Comput. Vis. Image Underst. 2010, 114, 641–651.

52. Cucchiara, R.; Fornaciari, M.; Haider, R.; Mandreoli, F.; Prati, A. Identification of intruders in groups of people using cameras and RFIDs. In Proceedings of the ACM/IEEE International Conference on Distributed Smart Cameras (ICDSC), Ghent, Belgium, 22–25 August 2011.

53. Foster, K.R.; Jaeger, J. RFID inside. IEEE Spectrum, 1 March 2007. Available online: http://spectrum.ieee.org/computing/hardware/rfid-inside (accessed on 10 June 2015).

54. Matic, A.; Osmani, V.; Mayora, O. RFID-based system for tracking people: Approaches to tagging demented patients.Lect. Notes Inst. Comput. Sci. Soc. Inform. Telecommun. Eng. 2011, 70, 60–65.

55. Xiong, Z. Hybrid WSN and RFID indoor positioning and tracking system. EURASIP J. Embed. Syst. 2013, 2013, 6.

56. Hsu, C.-C.; Chen, J.-H. A novel sensor-assisted RFID-based indoor tracking system for the elderly living alone. Sensors2011, 11, 10094–10113. [PubMed]

57. Hui, F.C.P.; Chan, H.C.B.; Fung, S.H. RFID-based location tracking system using a peer-to-peer network architecture. In Proceedings of the International MultiConference of Engineers and Computer Scientists, Hong Kong, China, 12–14 March 2014.

58. Lin, X.; Lu, R.; Kwan, D.; Shen, X. REACT: An RFID-based privacy-preserving children tracking scheme for large amusement parks. Comput. Netw. 2010, 54, 2744–2755.

59. Steffen, R.; Preißinger, J.; Schöllermann, T.; Müller, A.; Schnabel, I. Near field communication (NFC) in an automotive environment. In Proceedings of the 2nd International Workshop on Near Field Communication, Monaco, 20 April 2010; pp. 15–20.

60. Firence Card. Available online: http://www.firenzecard.it/?lang=en (accessed on 9 June 2015).

61. ExxonMobil's Speed-Pass. Available online: http://www.speedpass.com (accessed on 9 June 2015).

62. Washington Metropolitan Area Transit Authority. Shuttle Services at Metro Facilities Report; WMATA: Washington, DC, USA, 2011.

63. Collins, J. Gift Cards Go Contactless. Available online: http://www.rfidjournal.com/articles/view?762 (accessed on 9 June 2015).

64. Charette, R. Wal-mart to track clothing with RFID Tags. IEEE Spectrum, 27 July 2010. Available online: http://spectrum. ieee.

org/riskfactor/computing/it/walmart-to-track-clothing-with-rfid-tags (accessed on 9 June 2015).

65. Marks & Spencer Embraces Change. RFID J. 2014. Available online: https://www.rfidjournal.com/purchase-access?type=Article&id=11952&r=%2Farticles%2Fview%3F11952 (accessed on 9 June 2015).

66. Zara Builds Its Business around RFID. The Wall Street Journal, 16 September 2014. Available online: http://www.wsj.com/articles/at-zara-fast-fashion-meets-smarter-inventory-1410884519 (accessed on 9 June 2015).

67. Special NRF Report: Decathlon Develops RFID to Revolutionise Stores and Back of House. The Retail Bulletin, 17 January 2014. Available online: http://www.theretailbulletin.com/ news/special_nrf_reportdecathlon_develops_rfid_to_revolutionise_stores_and_back_of_house_17-01-14/ (accessed on 9 June 2015).

68. Gilart-Iglesias, V.; Maciá-Pérez, F.; Marcos-Jorquera, D.; Mora-Gimeno, F.J. Industrial machines as a service: Modelling industrial machinery processes. In Proceedings of the 5th IEEE International Conference on Industrial Informatics, Vienna, Austria, 23–27 June 2007; pp. 23–27.

69. Andújar-Montoya, M.D.; Gilart-Iglesias, V.; Montoyo, A.; Marcos-Jorquera, D. A construction management framework for mass customisation in traditional construction. Sustainability 2015, 7, 5182–5210.

70. Eriksson, H.E.; Penker, M. Business Modelling with UML; John Wiley & Sons: Hoboken, NJ, USA, 2001.

71. García-Chamizo, J.M.; Mora-Pascual, J.; Mora-Mora, H.; Signes-Pont, M.T. Calculation methodology for flexible arithmetic processing. In Proceedings of the IFIP International Conference on Very Large Scale Integration (VLSI-SOC), Darmstadt, Germany, 1–3 December 2003; pp. 350–355.

72. Mora-Mora, H.; Mora-Pascual, J.; García-Chamizo, J.M.; Jimeno-Morenilla, A. Real-time arithmetic unit. Real Time Syst. 2006, 34, 53–79.

73. Mora-Mora, H.; Mora-Pascual, J.; Signes-Pont, M.T.; Romero, J.L.S. Mathematical model of stored logic based computation. Math. Comput. Model. 2010, 52, 1243–1250.

74. Paul, A.S.; Wan, E.A. Wi-Fi based indoor localization and tracking using sigma-point Kalman filtering methods. In Proceedings of the IEEE/ION Position, Location and Navigation Symposium, Monterey, CA, USA, 5–8 May 2008; pp. 646–659.

75. Dwoskin, E. What secrets your phone is sharing about you businesses use sensors to track customers, build shopper profiles. The Wall Street Journal, 13 January 2014. Available online: http://www.wsj.com/articles/SB10001424052702303453004579290632128929194 (accessed on 9 June 2015).

76. Nikitin, P.V.; Rao, K.V.S. Performance limitations of passive UHF RFID systems. In Proceedings of the IEEE Antennas and Propagation Society International Symposium, Albuquerque, NM, USA, 9–14 July 2006; pp. 1011–1014.

77. Li, N.; Becerik-Gerber, B. Performance-based evaluation of RFID-based indoor location sensing solutions for the built environment. Adv. Eng. Inform. 2011, 25, 535–546.

78. Das, R.; Harrop, P. RFID Forecasts, Players and Opportunities 2014–2024; IDTechEx: Cambridge, MA, USA, July 2014. Available online: http://www.centrenational-rfid.com/docs/users/file/RFID_Forecasts_2014_2024.pdf (accessed on 9 June 2015).

79. Juels, A. RFID security and privacy: A research survey. IEEE J. Sel. Areas Commun. 2006, 24, 381–394.

80. Juels, A.; Rivest, R.L.; Szydlo, M. The blocker tag: Selective blocking of RFID tags for consumer privacy. In Proceedings of the ACM Conference on Computer and Communications Security, Washington, DC, USA, 27–30 October 2003; pp. 103–111.

81. RFIDControls. Available online: http://www.rfidcontrols.com/ (accessed on 9 June 2015).

82. Hohpe, G.; Woolf, B. Enterprise Integration Patterns, 1st ed.; Addison-Wesley: Boston, MA, USA, 2003.

83. Sivrikaya, F.; Yener, B. Time synchronization in sensor networks: A survey. IEEE Netw. 2004, 18, 45–50.

84. Zhang, F.; Deng, G.Y. Probabilistic time synchronization in wireless sensor networks. Int. Conf. Wirel. Commun. Netw. Mob. Comput. 2005, 2, 980–984.

85. Erl, T. SOA Design Patterns, 1st ed.; Prentice Hall: Upper Saddle River, NJ, USA, 2009.

86. Cubietruck. Available online: http://www.cubietruck.com/ (accessed on 9 June 2015).

87. Nikitin, P.V.; Rao, K.V.S. Antennas and propagation in UHF RFID systems. In Proceedings of the IEEE International Conference on RFID, Las Vegas, NV, USA, 16–17 April 2008; pp. 277–288.

88. Chen, Z.N.; Qing, X.; Chung, H.L. A universal UHF RFID reader antenna. IEEE Trans. Microw. Theory Tech. 2009, 57, 1275–1282.

INDEX

P

R

S

T

W